MW01641132

Taken for Grantedness

Taken for Grantedness

The Embedding of Mobile Communication into Society

Rich Ling

The MIT Press
Cambridge, Massachusetts
London, England

MIT Press books may be purchased at special quantity discounts for business or sales promotional use. For information, please email special_sales@mitpress.mit.edu or write to Special Sales Department, The MIT Press, 55 Hayward Street, Cambridge, MA 02142.

This book was set in Stone Sans and Stone Serif by the MIT Press. Printed and bound in the United States of America.

Library of Congress Cataloging-in-Publication Data

Ling, Richard Seyler.
Taken for grantedness : the embedding of mobile communication into society / Rich Ling.
p. cm.
Includes bibliographical references and index.
ISBN 978-0-262-01813-5 (hbk. : alk. paper)
1. Cell phones—Social aspects. 2. Mobile communication systems—Social aspects. 3. Interpersonal communication—Technological innovations—Social aspects. 4. Communication and culture. I. Title.
HE9713.L564 2012
303.48'33—dc23
2012009355

10 9 8 7 6 5 4 3 2 1

Contents

Preface: Mobile Phone Balloons

In some ways, the mobile phone is disappearing. I say this knowing that every week many millions of people in India get a mobile phone subscription for the first time. I say it knowing that literally every teen in Denmark and Norway has one. The paradox is, however, that in some interesting ways it is disappearing from our sense of what is remarkable. Mobile communication is becoming embedded in society. It is expected. It is possible to see this, in some deeply refracted way, by looking at the sale of balloons during the national holiday of Norway, May 17.

On this day, the Norwegians celebrate intensely. Marking the nation's liberation from the Danes in 1814 and serendipitously from Germany in that same month in 1945, May 17 is a walloping celebration. People dress in the bunad (the nation's traditional costume, which can cost many thousands of dollars), flags are flown, bands play, the king and the royal family wave to crowds from the balcony of the palace, and there are parades. The day is also focused on the nation's children. The children march (actually marching sounds too martial; they walk) in parades behind their school bands. Games and activities are arranged and, at least according to my family's tradition, they get all the ice cream they can eat (given that in Norway May 17 can be cold and blustery, this is usually a safe promise on the parents' part).

To bring this back to balloons and mobile telephony, the children also often get balloons with the ice cream. As a sociologist who has followed this practice for more than two decades, it is interesting to watch the coming and going of balloon styles. Each year at least fifteen or twenty different types of balloons are sold by street vendors, reflecting, in some abstract way, a measure of the times. There are often rather generic characters such as unicorns, sea horses, exotic fish, or hearts. There are also gender-specific balloons that reflect popular culture. One year it might be a cartoon character (Pac-Man has been a common motif), an icon from a popular movie, or an improbable-looking car. If, for example, Lightning McQueen from the film *Cars* or Ariel from *The Little Mermaid* is popular one year, that type of balloon gets into the mix. From the perspective of the person selling the balloons, the point is to capture the imagination of the child and to pocket the money from the parent. If unicorns work, that's great. If the current buzz is all about *Angry Birds*, go for it.

About fifteen years ago, in the mid-1990s, I noticed mobile phone balloons being sold on May 17. Here was an object that had been picked out of all the other possible artifacts of our daily lives and was being offered as an element in our children's celebration of May 17. And why not? The mobile phone was new; it was becoming commonly available to ever younger teens (and certainly to the preteens and children who make up the main balloon market); it represented something interesting and exciting; it was an icon of the time. I have not done any sort of scientific study, but it seems that the popularity of the mobile phone as a May 17 balloon has come and gone. In the period when teens and indeed preteens were first getting mobile phones, a big buzz was associated with the device. Now, however, getting a phone is not novel. More than 80 percent of ten-year-old Norwegian children have a mobile phone. From the perspective of balloon themes, mobile phones are no longer cutting edge. This year's selection included the standard fantasy figures, but I saw only one mobile phone balloon—probably a leftover from previous years' inventories.

This is admittedly a crude measure, and my methods of data collection are also crude. However, the idea that mobile phones have lost their buzz is only partially true. The latest iPhone or Android device is still a big deal. However, they seem to have lost their place in the iconography of balloons.

The "balloon index" gives us a small albeit imprecise data point. On the one hand, the mobile phone has become more central to the functioning of society; on the other hand, it is also becoming less worthy of notice. Perhaps the thing that is remarkable is that it has become taken as a given. It is remarkable that a communication technology has been adopted by billions of people worldwide and that these people literally send trillions of text messages. It is remarkable that the mobile phone has become intertwined in how we do commerce, interact with family, and coordinate our daily life and any number of tasks. It is central to how we interact with our partners, our friends, and our children. To be sure, great numbers of shiny new phones that promise the moon are available on the market. Nonetheless, the mobile phone has become, in some ways, unremarkable: It has become taken for granted. In this book I examine how this has taken place.

Thus, at one level, this book is about the "disappearance" of the mobile phone. This book is also describes what the mobile phone tells us about society. Much commentary on information technology focuses on what it will do for the individual. I want to think about what a technology does for society. Margaret Thatcher once said, "There is no such thing as society." I beg to differ. Thatcher was talking about the government's responsibility to individuals (she went on to say: "There are individual men and women, and there are families. And no government can do anything except through people, and people must look to themselves first"). There is such a thing as society. Following Emile Durkheim (1974, 14), we should consider social facts as things. Often they are "things" that are quite subtle. They are so interwoven into the flux of our daily lives that we take them for granted; we simply assume them of one another. We assume that others have clocks and can tell time; we assume that others have a mobile phone or that they can drive a car. How does this happen? How is it that we assume of one another the mastery of timekeeping? How is it that we have come to think it natural that we must make daily commutes (which have horrible side effects in the toll taken by accidents, wasted time, and pollution)? Why is it that we feel at a loss when we leave home without our mobile phone?

I argue that these are traces of our commitment to society. Thatcher was wrong when she said that there is no such thing as society. Indeed, to evoke Erving Goffman (1967), we experience our social nature whenever we interact with one another, whenever we move into a particular line of action vis-à-vis one another, and whenever we make reciprocal assumptions of one another.

In my previous book *New Tech, New Ties* (Ling 2008), I focused on the way that mobile phones are used to create and maintain social cohesion, drawing on Durkheim's notion of ritual interaction and effervescence as well as Goffman's and Randall Collins's (2004) insights into everyday interaction. These were the keys to understanding how we create social structures. I found that the mobile phone helps us to maintain and elaborate connections in our most immediate sphere of friends and family. We can call or text our spouse or our close friends with incidental news ("I had to drop off a project proposal; can you go to the store and get bread and something for dinner?") or as we arrange a face-to-face meeting ("Let's meet at Jazz Café off Lille Grensen at 11:00"). Teens can call their parents in moments of need ("Dad, I crashed the car, I am ok, but there is a big ding in the bumper"), or we can call close friends at important moments ("Guess what! Sandra is engaged!"). It is through these mundane and not so mundane interactions that we develop and maintain our social networks. It is true, as I noted in my earlier book, that the ringing of the mobile phone can disrupt our engagement in the present moment. Nevertheless, having access to our nearest sphere of friends and family helps in the larger project of sustaining social cohesion.

This idea has not always fallen on fertile soil. The mobile phone is not a device that has been entirely well received into our midst. Perhaps rightly, it is often seen as a bother, a distraction, or a waste of money. It rings at inopportune moments, and using it in public is often seen as a sign of vulgarity. In some situations it is downright dangerous (driving while calling or texting is a case in point). Beyond this, it is important to understand how the mobile phone is embedded in our daily lives. It is important to consider how it is restructuring social interaction. After all, it is by far the most pervasive of all information and communication technologies in the world. There is no other communication technology that has spread as fast or as far as the mobile phone. To fail to study the social turbulence it occasions is to ignore a major social change.

The people who populate our intimate sphere are vastly important to us. We rely on one another for emotional support and social interaction. It is this group who laughs with us in good times and helps us through the bad times. They give us good—and sometimes not so good—advice. They help pick us up when we are down and to see the "angels of our better nature" when other emotions might be in danger of gaining the upper hand. More than anything else, at a completely mundane level, our intimate sphere of friends and family helps us in getting through the day. Teens and young adults meet up with friends for a pick-up game or to study together. Couples touch base during the day to find out what they will have for dinner or to simply say "hi." We prepare food and eat together. We wash one another's clothes and clean up after one another (some do more of this than others, but no genders will be mentioned here). In addition, we run errands, we try to keep track of our common projects—think children—and we try to be there for one another. The mobile phone also allows us to be in contact when there is an immediate need. We saw this in the wake of the July 22 bombing in Oslo. A common theme for Oslo residents was that they almost immediately called or texted their closest friend or relative. They wanted to check on their status or they simply wanted to alert them and share their questions, fears, and concerns.

It is in these mundane (and sometimes not so mundane) but vitally important ways that the mobile phone has become so important. It may not have the same panache as it once did (we don't see as many mobile phone balloons, for example). Nonetheless, it gives us a constant link to those three or four other people who are so important in our daily lives. It is an essential tool in our maintenance of the intimate sphere. The internet also does this; indeed, our spouse or best friend is probably also on our "friends list" of whatever social networking site we use. However, dozens if not hundreds or thousands of other people who are not nearly as close are also on that list.

The mobile phone is different in this way. It is an instrument of the intimate sphere. It is our link to those people with whom we are closest. Indeed, when we do not have our mobile phone with us, we have in some small way betrayed the link with our intimate sphere. We might not be there when the gang is deciding on which café to meet at, or we might not get the message from our spouse that our child is sick and needs to be picked up at school. In other words, group coordination is hindered if people are without their phones. Beyond that, we might not be reachable should there be a real emergency.

Thus, I suggest that the mobile phone has become taken for granted. It has become woven into our expectations of one another. I suggest that the mobile phone, like several other technologies such as mechanical clocks, the automobile, and the internet, is a technology that has become a part of the social fabric. It is part of what holds us together. We use it often and we expect the same of others. We feel slighted when others do not respect the time, underestimate problems of transportation, or forget important information. We have, in other words, an expectation of reciprocity with regard to these technologies.

The expectation that others in our immediate sphere are continually available, regardless of where they are, is something relatively new. It has led to an extended sense of our position in the social group. We are led to contact others at the least provocation. Instead of saving up our daily tidbits, we text one another immediately, and in this way, some of us are engaged in an ongoing conversation. Christian Licoppe (2004) calls this "connected presence." We can always get in touch with each other, and indeed we are never really out of contact. Further, we expect our interlocutor to also be there. If, for some reason, he or she does not respond to our texts, we might suspect that something is amiss.

We have been fascinated with the mobile phone as it has seemingly invaded every sphere of our lives. The device has awakened our indignation and has been seen as a technological fetish. We join the queue when the latest smartphone goes on sale, but we also have learned to put it in a certain perspective. More than anything else, the mobile phone is increasingly a vital part of our being social. However, it is disappearing into taken for grantedness.

Acknowledgments

There are many people to thank for their help on this book. I want to thank Amanda Lenhart from the Pew Internet and American Life Project for her work on the focus groups discussed here. I also want to thank Scott Campbell of the University of Michigan for his work on the same interviews and for being a great colleague (in spite of his supporting Nebraska). Special thanks to Klaus Bruhn Jensen from Copenhagen University, who read an early version of the manuscript. I also want to give special thanks to Nalini Kotamraju and Leslie Haddon for their open-hearted comments that really caused me to focus on some of my (ill-thought-out) arguments. Thanks also go to my Mobile Communication class of fall 2011 who read the manuscript and helped me to understand the growing importance of mobile-based social networking.

I want to particularly thank Jonathan Donner for his suggestions. Indeed it was in a discussion with him that I got the germ of an idea that resulted in this book. If Jonathan gave me the germ of an idea, Jim Katz gave me an elegant formulation. Thanks to him for that. Over the years, my Telenor colleague John Willy Bakke has been a source of stimulating comments. My discussions with him about this book have been no exception. I also want to thank Gerard Goggen, Geoff Canwright, and Christian Licoppe for their comments and their collegiality.

Ralph Schroeder from the Oxford Internet Institute has helped me focus my thoughts around Durkheim and Weber, and Knut Holtan Sørensen of NTNU in Trondheim has been the source of helpful comments on domestication theory. Katrin Verclas of Mobile Active has been helpful in sharing her knowledge of mobile telephony in developing countries.

I want to thank my colleagues at the IT University of Copenhagen who have helped me by reading chapters or by letting me bounce my half-baked ideas off them. These people include Troels Bertel, Lisbeth Klastrup, Irina Shklovski, Gitte Stald, and Bjarki Valtysson. Thanks as well to the anonymous reviewers who gave me tough but fair feedback. Finally, thanks to Doug Sery, Judith Feldmann, and the people at the MIT Press for their support.

1 The Forgotten Mobile Phone

My sister-in-law, a librarian in a small town in western Norway, was staying with us while she attended a book exhibition in Oslo. On a pleasant Saturday morning, she and I took the train together into the city center. She was on her way to the exhibition and I was meeting my friends Per and Geoff for a cup of coffee. After boarding the tram, she started to look through her pockets and her small backpack with a quizzical look. As she did this, she told me that she had forgotten her mobile phone. She had a loose agreement with a friend to meet at the exhibition, and then she had plans to meet yet another friend for dinner. There were no concrete agreements as to time and place, only tentative arrangements. All three had assumed that they would call one another as the day progressed to agree on when and where they would meet. Not having a mobile phone put both of these plans in danger, since there was now no easy way to firm them up.

She considered turning back to get her phone but that would have meant a considerable delay, given the need to wait for a new tram, make transfers, and so on. I offered to let her make a call on my phone, but the others' numbers were only recorded in her own phone. I suggested that she could find a phone booth at the exhibition, call my wife for the phone numbers, and work out her lunch meeting. Another alternative was to count on the chance that she would meet the first friend at the exhibition. If she could meet up with the first friend, she could borrow her phone and call the second friend to firm up the dinner date. This was how we left it when we parted.

Our reliance on mobile communication becomes especially obvious when we find ourselves without it. According to Anthony Elliott and John Urry, not to have a mobile communication device is to be "walking blind, disconnected from just-in-time information on where and when you are in the social networks of time and space" (Elliott and Urry 2010, 61). We live in a web of interactions and expectations that are increasingly mediated through mobile communication. If for one reason or another we are without a mobile phone, we—and those with whom we interact—have to think up "work arounds." There are alternatives (finding one of the diminishing number of phone booths, borrowing a phone from a friend, banking on the chance that you bump into your friends), but these are all difficult and risky. The easiest thing is simply to have the phone with you.

At the same time, the mobile phone is seen by many as an annoyance. We are disturbed by its ringing. It causes stress. People note that life was so much simpler without it. To be sure, we can come up with a long list of reasons why the mobile phone is evil. But it is increasingly a necessary one, as shown by my sister-in-law's experience. In spite of all our protests, the mobile phone is nearly ubiquitous in many countries; there are billions of subscriptions in the world. Why the paradox? Why do we join in the chorus of people who say they hate mobile phones when we have them ourselves all the while? Some suggest that the mobile phone ties us into broad social movements and enables our participation in global society. This may be true, but I do not see it as the primary reason for the popularity of this necessary evil. The mobile phone has become important mainly because it facilitates the mundane aspects of our lives. It helps us to keep in contact with family and friends. It helps us to coordinate informal social interaction. To be sure, it is a tool of industry and commerce. In the guise of a smartphone[1] or a tablet, it gives us access, via "apps," to information and functionality beyond our imagination. But it is, perhaps more than anything else, also an instrument of everyday interaction. Because of the mobile phone, we are increasingly obliged by our social network to be available to one another.

We may often feel inclined to think of the mobile phone as a personal technology. We choose a model that expresses our sense of identity. It is something that we have on our person—just like, for example, our jewelry or clothes. It is a place where we store personal information and perhaps our favorite songs and photos. As with our jewelry and clothes, we consider the style, color, and functions of the mobile phone quite carefully before purchasing (Fortunati 2005). Further, we are to some degree judged on the style and vintage of our mobile phones (Haddon 2003).

However, treating the mobile phone as a piece of jewelry or a bit of clothing captures only a part of its effect on society. It is, after all, much more than a piece of "bling." It is perhaps just as correct to think of it, at least in its phase as a communication tool, as social technology. We are expected to be available to family and friends via the mobile phone. It is a device that embeds us in our web of social interactions. It provides us with individual addressability so that we can call individuals, not places (Sundsøy et al.

forthcoming; Goggin 2009; Ling and Donner 2009). It helps us to micro-coordinate our daily interactions and to maintain our social relations (Ling 2002). We use it to coordinate meetings and to send greetings. We send quick text messages to our buddies wishing them a happy birthday and to iteratively decide at which bar we will meet to uncork the celebration. We use it to make dentist appointments and to tell work colleagues we will be five minutes late because of traffic. We use it to coordinate our individual affairs in the flux of everyday life and we use it to exchange expressive calls and text messages with those in our intimate sphere. In this way, it is becoming a part of the broader social metabolism. Mobile communication gives us an individualized communication device that we use for both the large- and small-scale interactions that form the fabric of everyday life.

As a result, the mobile phone has become a part of the common expectations we have of one another. It is no longer just "nice to have." It is not used just for our own convenience. Indeed, the use of mobile phones by a critical mass of people facilitates the smooth functioning of everyday life. We take it for granted that we are always able to a call a spouse or a good friend, and conversely, they expect that we are always available. If we forget our phone or are without it for some reason, it throws a cog into our social interactions (Lasén 2011). As my sister-in-law discovered, we are starting to assume that those with whom we wish to interact also have a mobile phone as a part of their everyday kit. If *they* are not available via the mobile phone, then it becomes *our* problem. Originally suggested by James Katz, this might be termed the "Katz principle" (Weiner 2007; Katz 2008).We need to somehow work around those individuals who are not available via mobile communication and use other, perhaps less efficient, forms of communication with them.

We might well ask why this matters. Why should we think about the intertwining of mobile communication in our lives? Why should others be available whenever we want to call? As Jeffrey Hall and Nancy Baym suggest, isn't this a case of overdependence (Hall and Baym 2011)? Wouldn't we be better off without?

The embedding of mobile communication in society matters because coordination is increasingly done via the mobile phone. As the mobile internet gains pace (as it is in many countries), we will be further reliant on mobile access to information. At a personal level, knowing how the mobile phone is increasingly interwoven into society helps us understand why we have become so attached to it and why we can have a sense of anxiety (some would say freedom) when we are without it. Understanding the process of social embedding helps us to think about whether this is pathology or just a healthy need for social interaction. It helps explain our worry when coverage is poor or when there is a technical problem in the system. Most of the time these are just simple annoyances, since the mobile phone is mostly used for mundane interactions such as deciding when and where to meet for a chat or to decide who will shop for groceries and who will pick up the kids at school. But other times the need is more urgent. Thinking of the mobile phone as a socially embedded system helps us see why we become so concerned when we cannot use one to reach our child or our elderly parent. Taking this to the horrific extreme, it helps us to see why the commentary regarding the shootings on Utøya in July 2011 often circled around mobile phone contact between parent and child. In short, it helps us to understand our deep-seated need to have a communication channel to our closest family and friends. Once we understand this insight, it may help us to frame policies that take into account the needs of users as we move on to ever new generations of mobile communication devices.

Mobile communication is not the only technology that is socially ingrained. Other technologies are similarly rooted and facilitate the functioning of social interaction. In addition to mobile phones, we can think of the car and mechanical timekeeping[2] as technologies that have become embedded in our lives, or at least, in the case of the car, in the lives of people living in the vast swathes of suburbia. These technologies challenged existing systems and firmly established themselves as necessary to the functioning of society. The car, the clock, and the mobile phone have all become central tools in our everyday lives. Although obvious exceptions do exist, these technologies are unquestionably central elements of contemporary society. Given the separation between home and work, and the fact that, for example, children's activities are in one quarter of the city, our shopping and commercial needs are serviced in another, and our social lives are carried out in yet a third location, communication, coordination, and transportation (and the technologies that facilitate these) all have central importance to the normal functioning of our lives.

Looking specifically at the automobile, in the 100 years following the late 1800s, it slowly developed from being an odd contraption on the edge of society to being taken for granted. In the late 1800s, none of the major elements of today's automobile culture was in place. Cars were rickety contrivances. They rarely had cabs for passengers, they needed constant prodding and maintenance, and they were more often than not seen as the hobby of determined tinkerers or eccentric millionaires. As if to ensure cars' marginal status, the roads were poor, and there were few gas stations and even fewer repair shops (Urry 2007, 113). If you owned a car at this time, you owned one almost in spite of its lack of usefulness. Society was clearly oriented toward other forms of transportation. This had consequences for the way that people organized their lives. Work, shopping, and schooling were often within walking distance. Daily activities did not require the individual to move about to the degree that we often see today. Neither the automobile nor the culture of the automobile had yet gained the purchase that they have today.

If we fast-forward 130 years or so to today, we see a huge difference. For better or worse, the logic of an automobile-based society has firmly established itself. Parking lots, paved roads, service stations, and all the standard automobile-related features of life are widespread. The car has also spawned strip malls and shopping centers. It is often easier to drive a few hundred yards from one strip mall to another (and belch out the consequent pollution), since walking involves detouring around multilane streets that are more car- than pedestrian-friendly. In addition, a whole sector of society is oriented toward servicing the automobile and its drivers and passengers. There are not just "filling" stations but service areas where we can attend to the nutritional needs of both the car and ourselves and where we can buy music, kitschy art, and reading material. This is not something that any single individual has brought about; rather, it is the collective sediment of social activity. For the vast majority of people (and in particular those who populate the suburbs), automobile culture is a given; and, while it gives us freedom of movement, it also forms and constrains our social activity (Urry 2000, 190).

In addition to having reformed the urban landscape, the automobile has spawned a supporting ideology: Not only is the automobile experienced as an essential means of transport (Hjorthol 2000), it is also a part of our identity that inspires a sense of loyalty *qua* dependency resulting in bumper-sticker-like statements such as "You can have my new SUV when you peel my cold, dead hands off its leather steering wheel" (Greenhut 2003).

Moreover, we often tacitly assume that others with whom we interact have access to an automobile. When we plan social events, set up meetings, or simply engage in planning the mechanics of everyday life (shopping for food, seeking entertainment, or meeting up with friends), we tend to suppose that we and our friends and associates have a car. Here again the Katz principle applies. In many cities if our friends do not have a car we need to develop elaborate ways of working around the logistical problem. To the degree that others rely on us being at certain places, our not having a car becomes an issue not only for us, but for them.

The automobile—and the system of social organization it has spawned—is an established feature of many metropolitan areas. In some cases it is possible (and perhaps desirable) to use alternative ways of getting around such as public transportation, walking, or bicycling. However, for many millions of people, the only realistic option is the car. Getting to work on time, getting the children to their activities, and shopping all demand the use of the personal car. This is not the car-owner's fault; rather, it is the consequence of a social trajectory. It is a part of the social structure. The automobile and its accompanying transportation system set the ground rules as to how sociation takes place. It is not necessarily that the automobile facilitates more social interaction; indeed, the opposite is probably more correct. However, it is a technology that mediates how we carry out our social lives.

The story of mobile communication is much briefer than that of the automobile. Although various forms of mobile radio have been possible since the early 1900s, the popular adoption of the cellular-based mobile telephone system is more recent (Farley 2005). Like the automobile, until recently mobile communication was the province of either the rich or the technically resolute. Mobile phone devices were heavy and required inordinate amounts of power. They were quirky, and the coverage was spotty. From the mid-1990s onward, we have seen mobile communication become much more commercially accessible, and we have seen the rapid acceptance of the mobile phone. This arrived first in the developed world and now is growing in the global south. It is increasingly imposing its reciprocally expected logic on the daily life of small groups. Love them or hate them, mobile phones are things we ignore at the peril of causing extra work for the people in our intimate sphere. As my sister-in-law discovered, we risk derailing our social arrangements if we forget our mobile phone, since it is becoming a taken-for-granted part of daily life.

Technologies That Mediate Sociation

In this book I examine how the mobile phone facilitates social interaction. In addition, I look at it vis-à-vis other technologies that also perform the same service, in particular cars and mechanical timekeeping. To be sure, there are other technologies of this type. Internet-based technologies such as email, and in particular the web 2.0 technologies like social network sites, also mediate sociation and indeed are the focus of much research (Katz and Rice 2002; Husinger, Klastrup, and Allen 2010; Consalvo and Ess 2011). In many countries the boundary between PC-based and the mobile-based internet is becoming difficult to discern. The diffusion of smartphones gives us access to many web services, in addition to uniquely mobile phenomena such as ubiquitous access to one another and location-based services. We are also seeing the growth of pads and tablets that combine mobility and internet access in different ways.

The interpersonal interaction afforded by the mobile phone is a medium through which, most of the time, we reach out to individuals rather than groups. It offers an intimacy ("Hi beautiful, I enjoyed our date last night ;)") and also an immediacy ("Help! The car battery is dead and we have to pick up the kids at day care. Can you pick them up while I sort out the situation with the car? :("). Texting and voice interaction are the mainstays of this, though the rise of smartphones and the mobile internet is giving this type of interaction a new functionality. All of these together are changing how small groups and individuals interact.

To be sure, the mobile phone is no longer just a voice- and text-based person-to-person communication tool. It is increasingly giving us new ways of interacting with one another in groups by way of "friend lists" and the latest information resources. The above-mentioned web 2.0 sites, accessed increasingly via the mobile internet, are most often used for quasi-broadcasting or "mass self-communication" (Castells 2009). The melding of the mobile phone and the internet is bringing us new possibilities in this sphere. We use our mobile access devices to get information from "the cloud" and location-based data such as maps and route descriptions. In addition, our ever more advanced phones are becoming repositories of individual as well as shared information.

I am interested in understanding the social construction of practices using certain technologies that have become common enough to be a shared part of our collective lives. These practices have social facticity and, at some level, we are coerced to use them. As technologies and systems that aid in the development and maintenance of social interaction (Green 2002), they have changed how we carry out major social practices; they have also changed the physical arrangement of society at the expense of alternative approaches. Finally and most importantly, we have come to have a common expectation that others use these technologies in about the same way as we do ourselves.

Social mediation technologies are legitimated artifacts and systems governed by group-based reciprocal expectations that enable, but also set conditions for, the maintenance of our social sphere. The use of social mediation technologies is not simply a matter of personal choice; it is in general an assumed part of social interaction. We use these technologies to orient and organize ourselves for instrumental as well as expressive purposes. We use them in our roles as family member, friend, and colleague. They are a part of the baggage associated with group interaction, and as such they can form, mold, constrain, facilitate, and set the conditions for sociation.

As such, these technologies provide us with efficiency and utility. This is not to say that they are socially, economically, or environmentally benign, nor does it mean that no power dimensions are associated with their use. They can have serious consequences on these fronts. It is obvious, for example, that the car is far more polluting than the horse. In spite of this, we have adopted the car and marginalized horse-based transportation, and along the way we have also built a structure of justifications that further entrench our use of the car.

Technologies of social mediation may be the norm for broad swaths of society, as is the case with mechanical timekeeping. However, their use and adoption can be characteristic of smaller social groups (think, for example, of foursquare users). The web of mutual expectations does not only function at the broad social level. It may be that members of a particular segment of society, people living in one area or participants in a special institution, share the expectation that others in that group will use a particular mediation technology. The use of collective calendars and scheduling programs is an example of locally based social mediation technologies.

There are also limits to the reach of many social mediation technologies. Time and timekeeping is nearly universal. However, I do not argue, for example, that the car is essential for everyone. In many cases there are alternatives, and a person simply does not need one. Indeed, in Copenhagen, a bicycle or a bus pass is more useful in many cases. That said, there are also people for whom the car is essential. The same is true of the mobile phone. Many groups of people simply do not need one or have alternatives for interpersonal communication. At the same time, there are those who seemingly live and die by having their mobile phone with them. The mark of a social mediation technology is that, for a particular group, by agreeing to use it (or being forced to use it in spite of our principles), we facilitate the functioning of the group. To not use it marks us, at some level, as deviants, and indeed there can be social pressure associated with the use of these technologies (Markus 1987, 493). By acquiescing in their use, however, we support the broader interaction of the group.[3]

It is worth noting that I use the word mediation with some apprehension. The word mediation has a long history of being associated with communication. However, the word has a flexibility that allows us to use it in a broad spectrum of situations. Indeed, Sonia Livingstone entitled her Presidential Address to the International Communication Association "On the Mediation of Everything" (Livingstone 2008). As she notes, this term has been applied to any number of ideas, theories, and situations. In discussions of communication, for example, mediation is often seen as the encoding of information that allows its transfer and eventual decoding by an interlocutor. In this book, mediation is used to denote a technological intermediary that facilitates our social dealings. This plays on the idea that a medium can also be seen as something that intervenes between actors. I use it to focus attention on the tools that we use to carry out social interaction, particularly interaction with that part of our social sphere that is not immediately at hand. The devices we use to coordinate, communicate, and transport ourselves are thus mediating the social interaction. This is the basis for the selection of timekeeping, car-based transportation, and mobile communication as technologies of social mediation.

Interdependences of the Clock, Car, and Mobile Phone

The social impact of mechanical timekeeping and the imprint of "car culture" are clearly more profound than that of the mobile phone. In many countries, they are more thoroughly interwoven into the fabric of society. Nonetheless, the three technologies share similarities and synergies. By examining their diffusion we can trace how technical systems become embedded in society. These three examples show how we grow to have the mutual expectation that our social interaction will be mediated through certain technological systems.

In addition, the car, the clock, and the mobile phone have interdependencies (Urry 2007; Elliott and Urry 2010). The car and vast suburban spaces have changed the need for interpersonal coordination and made us more reliant on synchronization with others; hence the growing use of the mobile phone for coordinating our activities (Ling 2004; Carey 1988). Microcoordination is changing the role of timekeeping in society just as the mobile phone can be said to complete the automobile revolution in the sense that it allows for ad hoc coordination in urban spaces (Ling 2004; Townsend 2000; Kwan 2006; Licoppe 2007). Until the 1990s, small-group sociation in suburban settings relied on transportation to specifically agreed-upon places at specific times. These arrangements may have been made telephonically via landline phones, but once made, they were largely fixed. There was little chance to renegotiate the agreement. If there was a misunderstanding, or if some were delayed in getting to their destination, the others were simply left to wonder.

Mobile communication changes this dynamic. Rather than making an immutable agreement, we work out a meeting iteratively. This gives us the chance to fine-tune our social arrangements, since we can make allowances for traffic and the accessibility of different locations. Our use of the mobile phone is reconstituting our use of mechanical timekeeping in making and keeping appointments just as the use of the automobile has. However, clock time retains its role as the governing metric of many social situations—we cannot run airlines or large "just in time" manufacturing concerns on the type of negotiated microcoordination that teens use when agreeing on when to meet. In addition to these real-time negotiations, it is fair to say that the internet (and the mobile internet) is a part of this complex. Millions of amateur soccer and baseball teams, sewing groups, and antique auto clubs post their meeting schedules on the internet. Thus, the internet is a part of the complex of coordination, communication, and transportation that is in some cases being accessed via smartphones and tablets.

Standalone Artifacts

If there are technologies that facilitate sociation, there are also technologies that have no noticeable networking characteristics. They are perhaps better characterized as "standalone" artifacts. Examples might be a favorite coffee cup, a toothbrush, or an umbrella. There are obviously social aspects to the use of these devices. If I do not brush my teeth regularly, it will affect my social life. People might not stand as close to me, and they might make comments behind my back. On the whole, however, these artifacts are not central to the mediation of society. Indeed, most of the things I own and use can be seen as standalone artifacts. These things help me through the day and perhaps flatter my sense of style. My clothes keep me warm and communicate a type of status aspiration or affinity to others (Davis 1985). The furniture I use has concrete functions, and at the same time, it reflects my family's sense of appropriate interior fashion and the limitations of our budget. Some standalone objects, once they have been placed in the room, need an occasional fluffing or cleaning in order to be maintained, but they are only marginally involved in the mediation of my social interaction. Standalone technologies may offer convenience and provide a service, but they do not usually mediate interaction. They are not directly essential in the arbitration of my social life.

It is clear that we can reinterpret artifacts. It is possible to take the most banal technology, a nail file or paper clip, and use it as a medium for social interaction ("If I leave the nail file on my desk, it means we can meet at the bar tonight"). While this is possible, it is a radical rereading of their design, just as using a mobile phone as a paperweight is a rereading of its design. In each case it is a stretch to reinterpret the artifact, but it is also obvious that this is a matter of continuum and not black-and-white categories.

This is not to say that standalone technologies cannot have a disproportionate social impact, but this does not make them social mediation technologies. Think, for example, of the refrigerator (Robinson 1990; Levinson 2011). Refrigeration undeniably brought about social changes. It changed the way we store and prepare food. It changed our sense of nutrition, and it had a hand in changing the production and distribution of food. However, it is not a technology of social mediation. It is not a device through which we arrange and carry out significant social interaction. Thus, it is only marginally of interest to me if a meeting partner does not have a refrigerator (or a toothbrush or a coffee cup or a ski rack). By contrast, if they do not respect the use of clock time, if they cannot drive themselves to a meeting, or if they are not available via the phone, it can, following the Katz principle, become a problem for me.

Mobile Communication as a Moving Target

Finally, a caveat with regard mobile telephony: mobile communication is a central focus of this book, but it is very much a moving target. The functionality of mobile phones is not stable. The mobile phone has morphed into a device that encompasses a camera, a music player, and any number of other functions and "apps." Increasingly, people are adopting advanced mobile devices that give them relatively unhindered mobile access to the internet. The mobile phone is becoming more like a personal computer. Rather than simply being a single-purpose device through which we can call or text one another, it is also becoming the locus of social networking sites, text processing, email, gaming, payment and banking services, and much more (Ling and Svanæs 2011). In some cases, these features are similar to the traditional point-to-point interaction of telephony and texting, which people can use to exchange instant-messaging (IM) comments (Walton and Donner 2009).[4] The quasi-broadcast features are also becoming more popular (e.g., in Facebook and Twitter, a message from a single user is sent to many others). The nature of the mobile device is also changing with the introduction of tablets and pads. Finally, alternative networks like Skype, Viber, WhatsApp, and many more are challenging the position of the traditional network operators such as Vodafone and AT&T. Thus, mobile communication is a dynamic area. The advanced mobile devices that support this are currently most common in developed countries. Indeed, most teenagers in Denmark have some kind of advanced smartphone, whereas if we look to Bangladesh or Pakistan, we find very few users with this type of device.

These changes mean in part that we are using the mobile phone for more than our private interpersonal interactions (e.g., "Hi, what should we have for dinner tonight?"); mobile communication is also including "status updates" sent out via Facebook to our social networking "friends" ("I saw the best film last night and then had a couple of beers with my friend Scott"). In the first case there is usually only a single interlocutor, whereas in the second it might include members of our closest sphere as well as others with whom we have less social commerce.

Organization of This Book

This book has four broad sections. They are:

1. General discussion of social mediation technologies (chapters 1 and 2);
2. Application of the framework to mechanical timekeeping and the car (chapters 3 and 4);
3. Application of the framework to mobile telephony (chapters 5 through 8); and
4. Synthesis (chapter 9).

In this and the following chapter I lay the framework for social mediation technologies. I discuss the need for a critical mass, a supporting ideology, changes in the social ecology that sustain the technology, and a web of mutual expectations regarding use.

Chapters 3 and 4 use historical information to examine this idea in the case of clock time (chapter 3) and the automobile (chapter 4). These cases are of interest because they illustrate the general process by which social mediation technologies can be diffused into society. They also have synergies with the mobile phone.

In chapters 5 through 8 I use material from interviews, focus groups, and quantitative analyses, starting in the mid-1970s, to explore how mobile communication has become embedded in society. In addition, I discuss quantitative analyses of ownership and use in other European countries and more recently the work supported by the Pew Internet and American Life Project.

Finally, in chapter 9 I bring together the analysis of the car, the clock, and the phone to see how the relationship between these technologies can be seen as part of the same complex. The ability to coordinate and transport ourselves and our wares relies on these technologies. The car, timekeeping, and the mobile phone are a part of the warp and weft of society.

2 DeWitt Clinton's "Grand Salute" versus Technologies of Social Mediation

On October 26, 1825, near Buffalo, New York, Governor DeWitt Clinton officially opened the Erie Canal. To mark the event, Governor Clinton and his party set out from Buffalo to Albany aboard the packet boat *Seneca Chief*. Included in their baggage were two casks of water from Lake Erie that would be ceremoniously emptied into the Hudson River 363 miles away. As they set out, they received a "Grand Salute" from cannons placed along the route of the canal and further along the 200 miles from Albany along the Hudson, past New York City to Sandy Hook on the Atlantic Coast. The cannons fired in turn, one after another, along the entire route. Upon hearing the report of their neighboring cannon, a particular crew would fire theirs, prompting the next crew down the line. The fusillade, starting in Buffalo and going to Sandy Hook and then returning to Buffalo, took 3 hours and 20 minutes (Arthur and Polak 2006; Howe 2007, 222). That is, the signal traveled at approximately 240 miles per hour. By comparison, the Pony Express averaged worse than 10 miles per hour. As is obvious, the salute was highly labor intensive and was able to communicate only a simple bit of information, namely, that Governor Clinton and his two casks of water were on their way.[1]

For its time, this was an amazing feat of communication, albeit one that was not practical on a daily basis. Such efforts were reserved for special occasions. Indeed, aside from written messages carried by horse, boat, or on foot, not many ways of sending information were available to the average person.[2]

Another system used at about the same time was the optical telegraph. This technology had been developed in several European countries, most notably in France (Dilhac n.d.; Gleick 2011). The system used a series of masts placed on hilltops within sight of one another. Each mast had a set of flaps or arms, whose arrangement could be changed using a series of ropes and pulleys. Like naval semaphores, the right arm up and the left one down might mean the letter A, both up might mean B, and so on. The systems of coding were actually much more complex than this. In some cases, numbers were communicated that referred to specific actions in a codebook—for example, the semaphore for 1 meant attack, 2 meant retreat, and so on (Standage 1998). These systems allowed the communication of more nuanced information than did Clinton's series of cannons. Their utility, however, was limited. They could only be used in clear weather during the daytime. They required a large corps of workers to send messages along the route. The stations had to be located relatively near one another and with no trees in the way. In addition, the messages were subject to degradation. This meant that although the messages had much higher information content than those in Clinton's cannon system, they were also often subject to ambiguity and delays. At the height of the system, there were over 550 stations in France (Standage 1998). If conditions were optimal, messages could travel as fast as 300 miles per hour for short bursts.

Systems such as the optical telegraph and even more so the cannon fusillade were not used for interpersonal communication (Gleick 2011, 135; Standage 1998). People outside of the military, public administration, and some parts of commerce were forced to send traditional letters by post. All of this changed with the development of the telegraph by Samuel F. B. Morse (and others) in the 1830s. Rather than measuring the speed of messages in miles or kilometers per hour, the delivery was for all practical purposes instantaneous. In addition, arranging communication with distant individuals was far easier. There was no need for innumerable cannon crews, optical telegraph station personnel, or post riders (Standage 1998). Eventually, in the 1870s, at least in industrialized countries, the telephone replaced the telegraph. The telex (1930s) and the fax machine (commercially feasible in the 1970s) came along in turn, and these have largely been replaced by mobile phones, the internet, and texting. At each stage, communication technology has become increasingly democratized.

These various forms of communication show that development can take a variety of directions with a variety of outcomes. Countless communication technologies and systems have been developed. Many of them soon end up being discarded because of their own internal failings or because they do not fit into the general organization of society. In some cases, such as the fax machine, they enjoyed a short run of popularity[3] before succumbing to other technologies (Star and Bowker 2006). In fewer cases, they become so enmeshed in the routines of daily life as to become taken for granted. The telephone (and more recently the mobile phone), for example, has in many countries enjoyed widespread adoption. Along the way, this technology has made communication a part of everyday life. A person in Buffalo, New York, thinks nothing of calling his or her friend in Sandy Hook—or for that matter in Singapore or Stockholm—simply to have a chat. While there is an amazing infrastructure that supports the call, it is not an extraordinary event when it takes place. It may even be a problem when it does not take place. If the wayward son does not call home on Saturday or if the soon to be ex-girlfriend does not call back, trouble may be afoot (Lasén 2011).

The issue examined here is not how we harness great forces of labor and machinery to let the world know that two casks of water are on their way to the Hudson. The issue is how other, less cumbersome systems become embedded in society. How are they socially constructed, and how do they come to impinge on us? How do these technological systems gain their foothold? How do they become increasingly necessary, and how do they gain the "real but immaterial" (Richards 2009, 500) status of being the mutually assumed basis for mediated social interaction?

Domestication and the Adoption of Technology

Before we discuss the facticity of technologies at the social level, it is important to think about how we personally adopt and use them. There are several different approaches to understanding the diffusion of innovation at the individual level. One prominent approach is that described by Everett Rogers (1995) in his discussion of the diffusion of innovations. Rogers describes how innovative people are the first to adopt. These are followed by early adopters, the early majority, the late majority, and the laggards. Based on the response of the innovators and their interaction with the early adopters, the rapid adoption starts when the early adopters start to use the device. Following the illustrations that are commonly associated with Rogers, about one-sixth of the population is made up of the innovators; one sixth are early adopters; the early and the late majorities are about a third apiece; and the remaining sixth are the laggards.[4] This approach is widely used in commercial analyses.

Another approach, and indeed the one that is foundational for the approach in this book, is the domestication framework (Haddon 2003; Silverstone and Haddon 1996; Silverstone, Hirsch, and Morley 1992). This approach was developed in the early 1990s in the UK. Domestication theory has generally, though not exclusively, examined local/microsocial processes. The first focus was on the adoption and use of the PC (Silverstone, Hirsch, and Morley 1992), but the framework was later applied to mobile communication (Haddon 2003; Ling 2004) and, for example, the automobile (Sørensen and Østby 1995). This perspective has also been applied to the broader adoption patterns of media and technology (Berker et al. 2006). The expansion of domestication theory to phenomena outside the home describes what it means to bring these innovations into our daily routines.

Domestication theory describes how we work through the adoption, mastering, and use of different technologies. It considers how we are somehow captivated by the thought of having an iPad, a new car, a new color printer, a subscription to the *New York Times*, or a sofa. We then go through a mental exercise of how it would fit into our lives. If it is a physical object we might think about where we will place it, how we will wear it, or on which occasions we might use it. In a similar way, if it is a service, we will mentally place its use into our daily lives. Will we, for example, have the time to enjoy it? What will our family or our friends think?

At this initial adoption stage, we might follow the discussion of the object in question in the press, on the web, and among colleagues. We might seek out information about both its positive and negative sides and how it has been used by others. In addition, we are forced to think about how we would use the object. Why do we want it? Why is it important? What problems would it solve? What problems would it cause? What would others think of our having it, and how would spending money on it be justified to other family members? All of these questions and many more come into play.

At some point, we make the decision to purchase (or not to purchase) the item in question. If we decide to buy it, then it moves from being an abstract idea in our minds to being a real factor in our lives. Instead of just thinking about its use or placement, we really to have to find a physical space for it and really start using it. We are faced with the need to work it into our daily routines. We have to justify our purchase to housemates, we have to find the time to learn how to use it, and we have to budget for our use. It may clash with the intended arrangement of the home, and it may mean that we have to change our daily schedules so that it can be fully utilized. We have to perhaps displace other objects that we have in the home or in the office.

We saw these processes occur with the introduction of the TV in the 1950s. Each family went through a deliberation process of deciding whether to buy one. After it was purchased, it had to be wedged into homes that were designed and built before the conception of the TV. For example, the living rooms in many homes were designed with the idea that they would be used for polite conversation or listening to the radio. The adoption of the TV, however, meant that the arrangement of the furniture needed to be reworked in a more or less pleasing manner, and the din of the TV made conversation difficult (Gittu, Jørgensen, and Nørve 1985).

It is particularly at this early point in the cycle of domestication—the mastering phase—that clear elements of social shaping arise. The way that we use and display the device, object, or service gives us the opportunity to make our own imprint on the technology and its use (MacKenzie et al. 2001). When confronted with its actual use, we may have to jury-rig the way it is plugged in, or we may need to work out innovative ways of situating it in our homes. As the technology becomes established and the routines become fixed, the use practices begin to crystallize.

There are those cases where a particular type of consumption is imposed on us. For many people, group calendar systems and email are examples of this. In the case of email, while there were well-established routines associated with traditional paper-based information exchange, email was adopted by some and scorned by others. Eventually, however, an increasing volume of work-related interaction took place via email, and those who were not linked into the system presented a hindrance to the remainder of the organization (Markus 1987). Regardless of individuals' attitudes toward email, it became a factor in our lives. We had to master its demands and learn the process of sending and receiving material electronically. We had to make space on our desks for the computer and we needed to learn how to apportion time to reading and writing our email. The decision was not necessarily our personal wish; rather, it may have been imposed on us in an institutional setting where the efficiency of group was given priority over the individual.

Leslie Haddon (2006) notes that the domestication of objects and services into our lives is not a "one off" occurrence. Instead, we may settle on a particular arrangement for a time, but changing patterns of life and our changing situation mean that we might also revisit a particular solution. The TV that was the center of attention when first bought becomes antiquated and moved into the children's room. The PC that represented the pinnacle of sophistication becomes outdated. It is moved from the place of honor to a more remote position. And so it goes. There is a cycling of the objects and the services that we use.

In the final phase of domestication, we see the beginnings of the transition to facticity that is the concern of this book. In some ways, this discussion can be seen as an extension of the domestication approach. After we have brought the artifact or service into our life, mastered it, and started to use it on a daily basis, our consumption becomes a part of the way in which others evaluate us. The social reflexivity is the point here: Others construct their expectations of us, and we of them, in relation to our use of mediation artifacts.

The foundation for these expectations is put in place as we evaluate our own purchase of an object. When we are going through the mental experiment associated with the adoption of an object or service, one part of the evaluation is how others might see us or how we will use the artifact to interact with others. There is an essentially reflexive dimension to this. In performing this thought experiment we have to stand in the other person's shoes. We have to take the role of the other to imagine how they might see us. We ask ourselves if we would be seen as ahead of our time or sadly drab. Are we hopping onto the bandwagon too early or too late when we decide to buy a lava lamp, adopt email, learn the Charleston, or use Twitter? When we have indeed adopted the Charleston or Twitter, then we have to actually accept the judgment of others. We might be complimented for our foresight or our sense of fashion. We might gain the respect of those who are also considering adopting the technology. But the opposite possibility also exists. It might be that we are seen as gauche and vulgar (Fortunati 2003). We might be seen as being tragically out of touch with the times. There is a kind of *vox populi* directed at our consumption decisions. The display and use of the things that we consume helps others to better understand us. It gives them a sense of who we are and who we strive to be.

The cycle of domestication can be applied to a range of items. It has its basis in the study of information and communication technologies such as the PC, the internet, the TV, video recorder, and, outside the immediate context of the home, the mobile phone and the automobile (Berker et al. 2006). It gives us a lens through which we can better understand the personal decisions associated with the adoption of these technologies. The domestication approach can take into account the external pressures associated with the adoption of an item. It might be that *all* the people at work have a certain type of shoe or *all* guys in class have a particular sweater or model of telephone. This vox populi affects our individual consideration of which item to consume.

This discussion of domestication has taken the perspective of personal consumption. It is worth asking what happens when we move beyond individual or local adoption to wide-scale adoption of a technology. Moving beyond personal adoption, we need to consider how a device or system affects the functioning of the social group. How does the situation change when a technology or innovation reaches critical mass and moves from being something a few people have to being something everyone is expected to have? How does adoption function when the technology in question facilitates social interaction? Putting this into the vocabulary of Peter Berger and Thomas Luckmann (1967), how do technologies become institutionalized, and further, what is the social construction of this process?

Communication technologies bring up special issues, since their use assumes that others have also adopted them. Communication technologies are not "standalone" items. Rather, they are used to help mediate the interaction of the group. Nonadopters potentially become a "missing link" in the communication chain since they are not able to expedite the flow of information in the group.

The Facticity of Social Mediation Technologies

As noted in the last chapter, social mediation technologies are social constructions. They evolve from socially contextualized technical systems into socially embedded structures. As they pass from earlier to subsequent users, they can gain a sense of inevitability (Pierce 1960). The discussion of socially constructed phenomena has a long legacy in the social sciences. One strand of this discussion is the social constructionist perspective, which focuses on how social processes shape technology. This perspective looks at the formation of the artifact as one of co-construction (Bijker 1995; Campbell and Russo 2003; MacKenzie and Wajcman 1985). That being said, there is also an understanding of technologies as heavily scripted by both the nature of the technology and the commercial interests or power structures that promote them. For example, Leslie Haddon notes "ICTs come pre-formed with meanings through such processes as advertising, design and all the media discourses surrounding them" (Haddon 2003, 2).

We do indeed appropriate and adapt technologies to our own needs, and in this way, as Silverstone and Haddon (1996) suggest, we are active "co-creators" of technology. It is also true that technologies can, as Weber says, stiffen such that they restructure social interactions (Schroeder 2007). This suggests that although it is possible to socially shape technologies, as time goes on, the technologies are less open to "tinkering" (Wallis, Qiu, and Ling 2012). At the point of departure, many (though not all) technologies are pliable. The internet and the PC and their many applications are examples of this. With time, however, our collective acceptance of a particular technology constrains its malleability.

Beyond the shaping of technologies, it is important to think about how we shape our use practices and how the complex of technology and use practices become embedded in the very structure of social interaction (Star and Bowker 2006). Social practices are formed in an interaction with the technical systems. Our collective notions of how and when we are expected to use email, Facebook, or a fax machine (in its day) have an inertia that is difficult to avoid. Our common ideas of such practices gain the power to coerce our use (Durkheim 1938, 51). They take us by the ear if we are so unschooled as to resist them. We see elements of this in Durkheim's social facts, Weber's "iron cage," and Berger and Luckmann's institutions, which will we now look at in turn.

Social Facts

One of the most central discussions of how social phenomena are constructed is found in the work of the French sociologist Émile Durkheim (1858–1917). He described how the use of, for example, a particular language or a particular currency can "acquire a body, a tangible form, and constitute a reality in their own right, quite distinct from the facts which produce it" (Durkheim 1938, 7). These are social facts. He states: "I am not forced to speak French with my compatriots, nor to use the legal currency, but it is impossible for me to do otherwise" (ibid., 3). For Durkheim, there is nothing imaginary about these constructions. Rather, they have a "corporeal, palpable" facticity. They are real entities that do not go away, however much we might wish they would (Bauman 2005, 363). Durkheim writes that "the first and most fundamental rule is: *Consider social facts as things*" (emphasis in the original) (1938, 14). Society can only be completely understood by considering these phenomena.

Social facts are not under the control of single individuals; they are attributes of the society. They exist before we arrive on the scene and they exist outside our direct influence (Durkheim 1938, 51). They are developed and formed by slow accumulation into massive phenomena that are beyond the will of any single person. While they are a characteristic of the collective, their machinations impinge on us individually. It is also important to note that we have a tacit understanding of their existence and their role in society. Indeed, it might be said that we only really notice them when we choose to go against the current or, like Heidegger's hammer, when it fails.

It is clear that elements of Durkheim's social facts can change over time. Dialects of a language, for example, can change and shift over time. However, the sheer weight of history means that social facts are only marginally amenable to change. Further, we are constrained by social facts. If we ignore them, we place ourselves outside the common forms of interaction. Durkheim writes that social facts are imbued with a compelling and coercive power by virtue of which, whether he wishes it or not, they impose themselves upon him. . . . it asserts itself as soon as I try to resist. (Durkheim 1938, 51)

Further, he notes:

They come to each one of us from outside and can sweep us along in spite of ourselves. If perhaps I abandon myself to them I may not be conscious of the pressure that they are exerting upon me, but that pressure makes its presence felt immediately [when] I attempt to struggle against them. (Ibid., 53)

Durkheim uses the example of money. We could try to exist, for example, by bartering with goods, but this would conflict with the expectations of those with whom we interact. It is difficult if not impossible not to comply with social facts. By complying, however, we also support the cohesion of society. "The individual submits to society and this submission is the condition of his liberation" (Durkheim 1974, 74). When we do comply, we enjoy the efficiency provided by the common set of expectations.

Social facts become visible in the various rules and ideas of how society functions. They can also result in the establishment of institutions as well as forms of legitimation and power. Banking, for example, is an institution that is founded on the use of money. This, in turn, has resulted in a host of laws, institutions, careers, and stratagems for coming out on top.[5]

The Iron Cage

Durkheim gives us the basic idea that social processes can have facticity. It is also interesting to think about their evolution. It is here that a contemporary of Durkheim, the German sociologist Max Weber (1864–1920), is of interest. Weber developed a somewhat analogous concept, *stahlhartes Gehäuse*, translated variously as the "iron cage," the "shell as hard as steel," and "a steel-hard casing." Weber takes a more evolutionary approach than Durkheim. According to Weber, the notion of a calling is an example of the iron cage. The idea of a calling arose in the monastic orders, where it had certain flexibility. Eventually, however, it became a central organizing concept associated with industrial production. The general notion developed by the early Protestants has become hardened into a social imperative. Weber wrote:

The Puritan wanted to work in a calling; we are forced to do so. For when asceticism was carried out of the monastic cells into everyday life, and began to dominate world morality, it did its part in building the tremendous cosmos of the modern economic order. This order is now bound to the technical and economic conditions of machine production which today determine the lives of individuals who are born into this mechanism, not only those directly concerned with economic acquisition, with irresistible force. Perhaps it will so determine them until the last ton of fossilized coal is burnt. . . . The care for external goods should only lie on the shoulders of the "saint like a light cloak, which can be thrown aside at any moment." But fate decreed that the cloak should become an *iron cage*.

Since asceticism undertook to remodel the world and to work out its ideals in the world, material goods have gained and increasing and finally an inexorable power over the lives of men as at no previous period in history. (Weber 2002, 181; emphasis added)

While there are differences between Weber and Durkheim, some of the same elements are in place. Through a social process, a system of thought (a calling for Weber) or a social phenomenon (e.g., language and money for Durkheim) becomes part of the reified preexisting structure of society. Both constrain the individual, who has little choice but to accept the system and its consequences. Interestingly, Weberian ideas have been taken in the direction of understanding the consumption and use of social mediation technologies (Schroeder and Ling, forthcoming).[6] A common element for both Durkheim's social facts and Weber's iron cage is that once in place, they impinge on the flow of events. Their establishment as social phenomena means that they are not inert elements of the landscape. They have a hand in determining how situations play out.

Social Institutions

Weber and Durkheim speak at the broad social level. A meso-level approach is provided by the Austrian-born sociologist Peter Berger (b. 1929–) and the German sociologist Thomas Luckmann (b. 1927–) in their discussion of social institutions. As with both Durkheim's social facts and Weber's iron cage, the institution described by Berger and Luckmann is a social construction. Often the word "institution" is use to describe a large organization or structured pattern that attends to a particular portion of life such as education or religion. In the work of Berger and Luckmann it describes an interaction of individuals at the local level. They write:

Institutionalization occurs whenever there is a reciprocal typification of habitualized actions by types of actors. Put differently, any such typification is an institution. What must be stressed is the reciprocity of institutional typifications and the typicality of not only the actions but also the actors in institutions. The typifications of habitualized actions that constitute institutions are always shared ones. They are available to all the members of the particular social group in question, and the institution itself typifies individual actors as well as individual actions. (Berger and Luckmann 1967, 54)

As this "reciprocal typification" gains a history it becomes increasingly crystallized. When person A does action 1, person B observes this, interprets it as the initiation sequence, and counters by doing action 2. Over time this becomes mechanized into an interlaced sequence of events that are performed without anyone having to talk about it. Further, the people involved in the set of actions eventually construct justifications for this particular way of doing things. Diversion from this process may invalidate the result. As this interpersonal practice (along with the justifications for it being just so) is transferred to subsequent people, it becomes a social institution.

There is an evolutionary dimension to this process, and Berger and Luckmann's institutions are also scalable. They can apply to large or small social groups. Looking at the smaller end of the scale, John, for example, might start to make dinner. When his roommate Jim sees this, he has learned to start setting the table. With time, John and Jim's interactions become increasingly crystallized. John and Jim may even go so far as to develop justifications for this way of having dinner. It might be that John claims to be hungry first or that Jim is protective of the china or likes to have the silverware arranged in a certain way. The complex of the "reciprocal typifications" eventually becomes a system of interactions that seems natural and perhaps unquestionable. It might be that John occasionally wants to use chopsticks or that Jim would really like to prepare the cream of chicken soup he favors and not the sushi that John likes. Adjustments can be made to the system. However, the patterns become so entrenched that it is easier to accept the flow of interaction than to buck the trend.

If a third person, Jordan, enters onto the scene months or years after John and Jim have developed their way of doing things, he may not understand why it is happening that way. Why is it always John who makes the food and Jim who sets the table? John and Jim may have their justifications (or legitimations, as Berger and Luckmann would call them) ready at hand. It is often only after having to include new individuals into the nascent institution that the need to enunciate the justifications arises. If Jordan becomes a regular part of the dining group, he will not only have to learn the routine specified by the institution, he will also have to respect the legitimation structure.

The work of Berger and Luckmann allows for the examination of power structures. It is possible to consider why some participants are perpetually able to command the situation and others need to follow. It might be that John gets to determine the timing and the content of the meal and Jim and Jordan respect this order even when they would prefer an alternative.

They might rise up in revolt when John serves liver and broccoli casserole. However, barring any culinary insurrections, they may be willing to follow along and accept John as the gastronomic master of their home.

A local dinner "institution" can arise, and the same things can be seen in relation to the adoption of new technology. As with Weber's iron cage, the evolution of Berger and Luckmann's institutions includes a process of decreasing malleability. Social phenomena (our routines of using mobile phones, for example) form as a result of the "reciprocal typifications of habitualized actions" (Berger and Luckmann 1967, 135). As these actions are embroidered with legitimations and imbued with power structures, they become more stable and indeed embedded in society. This said, they are not quite the same as Durkheimian social facts that are so thoroughly interwoven into society that we are unable to imagine them not being there.

The Components of Social Mediation Technologies

Several elements characterize a social mediation technology as it becomes embedded in society. The technology or system needs to be widely diffused in the group or society such that there is a perception of critical mass, and there also needs to be a system of justifications or legitimations. The adoption of a new technology can challenge the established forms of practice, and it restructures the physical world as it develops into a reciprocally expected phenomena.

Diffusion and the Perception of Critical Mass

A technology of social mediation needs to be perceived as having a critical mass for a group of users.[7] The diffusion of social mediation technologies proceeds from including only the solitary the hobbyist or innovator through a transition into a more social phase. Our adoption can be the result of an individual decision or perhaps peer pressure. Alternatively, a device can be imposed by an employer. The boss, for example, might say that we have to have a mobile phone so that we are more accessible to customers. Indeed, this sort of scenario is often the case for early adopters (Johannesen 1981). It is also the case that many people buy a mobile phone for their elderly parents out of concern for their well-being. Rather than our making a personal decision, the technology is thrown into our lap. Once there, however, the domestication process proceeds. By dint of its usefulness (or by dint of our boss pushing it onto us), we work it into our routines.

In this more individual phase, the device or system is not integral to broader social processes; in fact it may exist in spite of them. It can be seen as a type of extravagance or diversion that may have yet to show its usefulness (it might also end up in a drawer gathering dust). It may gain a foothold if we think that it can make an existing process more efficient. In this sense, it can grow to occupy what economist Kenneth Boulding (1978) called an empty niche (discussed further below).

As the usefulness of social mediation devices becomes recognized we start to use them in our interaction with others. The device becomes increasingly embedded in our lives as others also adopt its use. With time we may develop the sense that there is a critical mass of users. It is important to note that this is a perception, and not a specific point in the diffusion process (Lin and Hsu 2009; Lou, Luo, and Strong 2000). It is difficult to say that the threshold is, say, 23 percent or 45 percent of others in a population who are users. Instead, we are dealing with the *perception* of critical mass. We may, for example, hesitate to adopt a group calendar system until we have the sense that a significant number of our colleagues have also jumped on board.[8] This perception can come from broader information campaigns run by our employer or from chats with colleagues (or more likely both). From whatever the source, we get the idea that we would not be alone were we to adopt. Indeed, there may be strong peer pressure to do so (Lou, Luo, and Strong 2000). If there is a critical mass of users and we do not adopt, we will be out of the loop, or worse, we will be a drag on the group. We will not be aware of key meetings. Our reluctance to use the calendar system may raise flags with our boss, and others will have to compensate for us (you are hearing, dear reader, the voice of experience).

The value of having a mediation technology increases for users as subsequent people also adopt. Those who have already adopted have an incentive to recruit other users in order to validate their decision. If only a few people update their calendar status or book meetings, the net-based calendar system is of little value. In the initial phases of diffusion it is the early adopters who pay the highest price. The hardy souls (or perhaps eccentrics) who try things out and work out the kinks get little value in the start, albeit they may have a stronger hand in the social shaping of processes. When the first problems are worked out and as the system gains utility, others may also be more likely to adopt. At this point, there can be a rapid transition period when non-users are recruited. There is a personal price to pay for adoption but there can be a group-level payoff. Commitment to use the online calendar, for example, means that we need to learn the system and spend some of our time filling in appointments. However, as the collective commitment reaches the critical threshold, there is an increasingly common idea that it is a valuable and indeed a necessary task, since it facilitates group interaction. To not use it somehow betrays the group.

As the group moves through the criticality zone, as the technology matures and becomes more refined, and as its social utility grows, pressure increases on the nonadopters. Users can exert informal pressure such as giving casual tutorials and unsolicited suggestions that serve to push the holdouts over into the camp of the adopters. The holdouts may get the sense that they are missing out on utility afforded by the system and have to continually deal with other's frustrated assumptions that they own or use the device in question.[9] At the same time, the alternatives may be degraded. The employer, for example, may no longer supply paper calendars (Lou, Luo, and Strong 2000).[10]

The increase in the general value of the system as additional users adopt the system is known as the issue of network externalities; that is, the individual buys the phone or the fax machine for personal utility but incidentally creates value for others who also have one. The same idea has also been formulated as "Metcalfe's law," where the value of, for example, a telecommunications network is proportional to the square of the number of connected nodes. Others have correctly noted that this formulation dramatically overshoots the mark (Briscoe, Odlyzko, and Tilly 2006). Additional users increase the value of the network, but if the utility for all users of the system truly went up with the square for every additional node in the system, the person in China or India who bought the six billionth mobile phone would make an implausibly large contribution to the total utility of the mobile communication system. This is clearly not the case. The central point, however, remains. There is greater utility to the group as the level of participation increases. In case of the telephone system, this is obvious. If only one person had a telephone, he or she would have no one to talk to. As a communication form becomes the medium of interaction, individuals are pressured to adopt the particular device, be it a mobile phone, a fax, an online calendar system, or instant messaging.

One example that shows the diffusion/critical mass issue quite clearly is the adoption of the fax machine. The technology had its roots well back into the nineteenth century,[11] but it only first became widely adopted in the late 1970s. From then until the rise of email and scanned documents, the fax machine was a key office technology. Fax machines were in millions of offices and it was (and sometimes still is) common to have a fax number along with a telephone number on business cards. According to Markus (1987) there was a social aspect of adoption. Adopters did not act alone; rather, they were influenced by those who had already adopted. Early users paid a relatively high price (both in cash and in trouble and toil) and received a low benefit in relation to those who came later.

As with the early landline telephone, the first people who adopted fax machines often bought them in pairs. Eventually, the functionality grew, the utility of the device became obvious, and the price fell to the degree that adoption began to take off. The fax machine fulfilled the expectations of the initial users, and based on this, they become missionaries (Markus 1987; Rice 1982). As adoption increased, the smaller local groups of fax users grew into an increasingly large seamless network. There were enough interlocking devices such that individuals with the need to send and receive documents could ask one another if they could "fax it." With time, the perception of a critical mass grew and thus pressure increased on the holdouts. These nonadopters began to have neighbors and colleagues from whom they could borrow a fax machine, but this could be cumbersome and could try the forbearance of the fax owner. In the UK, a postal strike in 1987 was, for many, the final straw. Growth in adoption of the device was perhaps steady, but the strike pushed fax ownership from being useful to being essential since it allowed people to send and receive letters and information in spite of the strike (Gunthorpe and Lyons 2004). The critical mass became self-sustaining. That is, those without a fax machine felt the pressure to adopt, and their adoption further increased the pressure on those who had not yet adopted. Those who were lagging in adopting the fax were, to a greater or lesser degree, coerced into buying one since their sluggishness was becoming a problem for others. In some ways the laggards' purchase of the fax machine provided more of a benefit to the broader community than they received themselves since it filled in the missing links in the faxing network (Markus 1987, 506). Interestingly, if the postal strike had come earlier in the diffusion process, before the fax had become established, it would not have caused the same effect.

As the example of the fax illustrates, some mediation systems become the standard only to be abandoned when they are out-competed by a newer technology. For example, among teens email is passé and Facebook or texting is current (at least at the time of this writing). However, while one form is dominant, it is difficult to dislodge it (Star and Bowker 2006). Further, not all products reach critical mass. In some cases, there is the beginning of diffusion, but then, for example, pricing changes may stunt its growth and the diffusion recedes. This can be seen in the case of video telephony in corporations (Sundsøy et al. forthcoming).

•

In the discussion up to now we've made the general assumption that all people in a population adopt a mediation form. However, it is worth noting that this need not be the case. Adoption of a particular form of social mediation can be limited to certain groups. Teens, not elderly people, for example, are the most common users of social networking sites in many countries.[12] The point is that a mediation form can reach critical mass within a particular group or even within a clique. It is interesting, for example, to see the way that some groups rely on mobile-based Facebook for coordination, while others use foursquare and yet others use texting.

Jonathan Donner and Shikoh Gitau have shown that certain mobile social networking sites such as MXit have become a common form of interaction among teens in South African townships (Donner and Gitau 2009). Teens started to use sites such as Facebook, or in the South African context MXit, to announce get-togethers, to exchange photos, and to work out the various dimensions of their social lives. Looking at another mediation form, teens are the dominant users of texting (Ling 2010; Ling, Bertel, and Sundsøy 2012), but they are more marginal users of email. For many teens, texting or Facebook is the forum in which to exchange their views on a recent party or to gossip about the new couple in their class. By contrast, there is the perception of a critical mass for email among middle-aged cubicle occupants as a forum for the exchange of office jokes. These social mediation technologies might be dominant in some groups while they are ignored in others. The notion of critical mass need not apply to a whole society, but it can apply to smaller sectors.

There is often the sense that reaching critical mass for a social mediation technology is a positive event. It can, however, have perverse effects. This can be seen in the case of automobiles. If I buy a car in order to be able to visit more friends, I can use it regardless of whether or not my friend has one. Unlike the externalities of the fax machine, there is no direct network effect of adding another car to the highways. However, seen from a broad social perspective, once the automobile has established itself, it imposes its logic on the physical arrangement of housing, jobs, schooling, shopping, and a variety of other institutions; we are, in effect, obliged to use the car. The very physical structure of society forces it on us. In the words of John Urry, "the car is immensely flexible and wholly coercive" (Urry 2000, 59). Once the critical mass of car users is reached, the logic of the car/suburban infrastructure imposes itself on the individual. The more the society is based on automobiles, the greater personal *dis*utility we experience by not having a car. This also has spillover to those with whom we interact socially. If I need to be chauffeured to and from all our common events, I become a drain on their forbearance.

Legitimation

As a technology of social mediation embeds itself in our practice, we construct justifications, or what Berger and Luckmann (1967) call legitimations, for (and against) its adoption. Berger and Luckmann say that our mutual agreement to use, for example, clock time "requires legitimation, that is, ways by which it can be 'explained' and justified" (1967, 79). As a technology begins to be perceived as reaching a critical mass, these justifications are used to help convince wavering colleagues to adopt, for example, the office's online calendar system. They are also used to defend the ownership and use of the technology by those who have already adopted it. It is not enough that a technology is useful or functional; it also needs to have a set of legitimations that facilitate its adoption. Thus a collective narration develops that supports (or may eventually hinder) adoption (Flichy 2007, 11).

To be sure, legitimations exist both for and against a particular technology, and these operate at both the interpersonal and the commercial levels. Advocates of the entrenched system use them to protect that position, while those who are in favor of the new technology seek out chinks in the armor that favor their approach. Following Berger and Luckmann, legitimation makes systems "objectively available and subjectively plausible" (1967, 110). Legitimation "explains" the institutional order and instructs us as to which actions are appropriate in a particular situation. More importantly it tells us why these actions are important.

The decisions regarding our use of technologies are, to some degree, determined by referring to broader social legitimation structures. These help us understand, for example, the reasons behind our use of the mobile phone and how these match other social values such as punctuality, efficiency, and sociation (Berger and Luckmann 1956, 112). At its broadest level, legitimation structures may integrate broad areas of meaning into a more or less unified totality. The people who are punctual, efficient in their work, and maintain a social network are model individuals.

When speaking of emergent technologies, we see an evolution of justifications. This is important when use of the social mediation technology moves beyond the interaction of the early (perhaps somewhat atypical) users to subsequent adopters. As it moves from the innovators who treat the benefits of the technology as self-evident to those who those who lag behind and who perhaps feel less pressure to adopt, the need to communicate the justifications becomes more important. The early adopters often fight against the entrenched reasons for the existing way of doing things. This campaign can be assisted by those who market the innovation. Indeed, their advertising content can provide the raw material for the construction of legitimation structures. Legitimations can also be a soft form of power used by those who are charged with the introduction of, for example, an unpopular online calendar system in a business. They can use many more-or-less transparent justifications in carrying out this errand.

The rise of a viable alternative can awaken a counter-legitimation by supporters of the original technology in order to defend against the heretical alternative. The jousting between alternative systems and their legitimation can be engineered based on the inherent characteristics of one system versus another. Telegraphy operators at first dismissed, then disparaged, and later desperately competed with the telephone (Standage 1998). If a person begins to stray from an established ideological system, he/she becomes problematic—or in the words of Geoffrey Star and Susan Leigh Boker (2006), the heretic becomes a "reverse salient"—to those who hold the faith. The people who stubbornly cling to sending letters in the post when faxing is so much faster present a problem to the users of fax machines. Those who cling to their paper calendars when the rest of the organization is online are seen as resisters who disrupt others' attempts to streamline coordination. Each side of the discussion attempts to keep followers in the fold using various elements from the canon of justifications. This can cause hard feelings and hinder interaction.

Legitimation systems can also outlive their usefulness:

> Habitualization and institutionalization in themselves limit the flexibility of human actions. Institutions tend to persist unless they become "problematic." Ultimate legitimations inevitably strengthen this tendency. . . . This means that institutions may persist even when, to an outside observer, they have lost their original functionality. (Berger and Luckmann 1967, 135)

This was seen, for example, in the period leading up to World War II, when officers in the U.S. Army had the sense that a horse-mounted cavalry was more efficient for ground-based military maneuvers than motorized forces (specifically the tank). This was clearly an absurd notion, but the strength of this ideological structure (along with a long history of mounted cavalry) was such that it was able to suppress other alternatives until long after its days were past. It was not until horses were shown to be hopelessly vulnerable to tanks that the colorful ideology of mounted cavalry was quietly dropped (Goodwin 1994, 52). This shows how legitimation structures can outlive their usefulness and also serve to mark power differences. Those who advocate one or the other side of the cavalry-versus-tanks debate or the post-versus-fax debate will end up controlling the purse strings and thus will also have the ability to advocate better for their side of the discussion.

For those who share a position, ideologies promote solidarity. The willingness to adhere, and also the willingness to police the adherence of others, can promote this solidarity. There can be moral dimensions to the use (or non-use) of these systems. For example, the people using the group calendar at work may share a collective sense of efficient coordination. Others will see the online system as being the handmaiden of the soulless devil. The seemingly innocuous barcode system is in this position. The barcode system allows for better control of merchandise and allows stores to track the flow of goods. However, on internet blogs there is a somewhat loopy campaign suggesting that barcodes contain the 666—the mark of the beast.[13] Thus, the technology is supported by many in the name of efficiency, and it is opposed by others based on supposed religious issues. It is clear, however, that the one has been trumped by the other.

Social Ecology

As a social mediation technology diffuses and thereby gains critical mass and legitimation, it can also change the social ecology (i.e., the physical structures or routines of society) in its own image.

The term "social ecology" has been used in different ways. It was first used by Harmon DeGraff (1926) to describe the rural-to-urban migration in the United States. It has grown to focus on the interrelationships between individual elements of society on the one hand, be they persons, institutions, or systems, and the environment in which they exist, on the other. It has been applied to the management of interorganizational relationships (Astley and Fombrun 1983). It has also been applied to the treating of adolescents with a history of abuse or behavioral problems and the complex of family, friends, neighbors, and teachers with whom the teens interact (Borduin, Schaeffer, and Heiblum 2009; Ennett et al. 2008). As one might expect, it has also been applied to the social aspects of pollution (Bookchin 1987).

Here I will use the idea of social ecology in a somewhat different sense. Specifically, it describes the physical structures and routines of society that are changed as social mediation technologies are adopted (see, e.g., Stokols 1992). Where legitimation exists in the sphere of attitudes, social ecology exists in the sphere of concrete structures and existing routines.

As a technical system gains a foothold in society it will begin to impose its own logic and rearrange the context into which it is diffusing. As a device goes through the diffusion process, it is possible to see the "space" that it moves into as an empty or an open niche (Boulding 1978). As it moves into a niche, it can out-compete other previously established technologies and reform the boundaries of the niche. This conception plays on the idea in biology where there are open niches that we occasionally stumble into. Invasive species such as rabbits in Australia, cats in Indonesia, and starlings, kidzu, and tamarix in North America, among numerous other examples, illustrate the often unfortunate results of open niches (Money and Hobbs 2000). They show, however, that there was clearly a niche for rabbits in Australia, for example, that went unfilled until some British fellows, intent on having rabbit stew, imported them from the home country. The rabbits liked the new arrangement and complied by rapidly expanding to fill the open niche. Along the way, they changed the conditions for other animals and plants. That is, they changed the boundary of the niche by outcompeting native species and thus readjusted the ecology of Australia. Another example is tamarix, a bush that came to the United States from the dryer parts of Africa and Asia. In the 1800s it was imported to the United States as an ornamental plant and used to fight soil erosion. Eventually it found its way to the rivers of the southwest, where it made itself at home and proceeded to out-compete other plants such as cottonwood trees and to suck up large amounts of scarce water to boot. Thus, tamarix and rabbits not only expanded into what was an open niche, they readjusted the position of other related elements in the ecology. The niche was not fixed, but rather it evolved to the benefit of the newcomer and at the expense of the preexisting ecosystem.

This line of thought comes from the work of the economist Kenneth Boulding, who used the metaphor of ecology when describing different artifacts such as autos, houses, and totem poles as populations (Boulding 1955). His point was that various commodities can be viewed as populations that inhabit positions (or niches) in the social (as opposed to the natural) landscape. Based on their position in the commercial or social firmament, they grow and, in the terms of marketing people, they gain market share. Eventually they reach some equilibrium population in a particular niche. The situation is, however, dynamic. Some commodities flourish for a time only to disappear—for example, eight-track tape players. Others are mutations that are somehow ill adapted for life from the start and quickly die out—for example, hip-pocket-sized vinyl record players.[14] There are other commodities, however, that flourish beyond expectations, just like starlings in the United States and rabbits in Australia.

As a species becomes common, it can change the dimensions of its niche. While it is common to see a niche as being rather strictly bounded (koala bears cannot exist where there are no eucalyptus trees, fish need water to breathe, etc.), it is also possible that a niche will expand and change its boundaries. This can happen when the occupant somehow out-competes neighboring species and thus co-opts their position. Technologies can alter the niche into which they move. As a technology moves into and dominates a niche, it can reconstruct the broader ecosystem. The fax machine, for example, reconfigured the ecology of sending paper-based letters. Previous to its widespread adoption we used the postal system or some other form of courier service. As the fax machine became popular, it changed (or at least threatened) the role of the post.[15]

As we will see, the social mediation technologies examined here all grew into a niche and competed against existing occupants in nearby niches (mobile phone vs. landline phone, car vs. horse, and mechanical clock vs. sundial). In this process, a particular device gains a foothold and shows its usefulness. As time goes on, resources are reallocated away from one system and toward another. In addition, habits of use are challenged and change. Issues of power are also worked out when political, commercial, and even informal power structures support one alternative over the other. Thus, the preexisting occupant of a niche is pushed aside by new occupants that impose their own logic. Social mediation technologies challenge and then change existing social structures. Our collective decisions to use them disrupt one regime and substitute another. The ancillary systems that were developed to support one system must also adapt as the new occupant arises.

Reciprocal Typification

Social ecology focuses on the way that a new technology will rearrange interrelationships between elements in a social context. In addition, technologies of social mediation rearrange our expectations of one another. Indeed, this is perhaps the core concept describing technologies of social mediation.

At some point in the adoption and diffusion process we stop being surprised at others' use of a technology. Somewhat later we might start to assume it of one another. In a sense, this progression means that the technology is becoming a tacit part of society. At some point there is the reciprocal assumption that others are also users. We take it for granted. When diffusion has reached this level, the social mediation system enables but also constrains interaction. It gives us the freedom to interact with other users, but it also demands that we are users. Nonusers are exiles.

To an increasing degree I assume that others will respond to my emails or come to meetings arranged in the group calendar. These assumptions regarding the mutual mastery of these social mediation technologies facilitate social interaction. Assumptions regarding access to technologies can indicate an individual's social status. The assumption, for example, that a person uses a particular technology does not necessarily hold for people at the fringes of society, either within a country or on a global basis. Indeed, a way that impoverishment might be felt is that a person or a family does not have access to items that others take for granted (Ehrenreich 2008; Horst and Miller 2005).

In some cases, I have to encourage (or push) another member of the gang to adopt one or another form of mediation (Donner and Gitau 2009). If they do not, and if the device or system gains currency, nonusers may be seen as deviants. It is the same with email and, in some groups, the use of instant messaging, Facebook, Google+, Pinterest, or Instacam. These technologies started out as being interesting or marginally useful, but at some critical moment in the context of the group, they became a reciprocally assumed part of the social interaction. When we can collectively make these types of assumptions regarding a technology, it is indeed embedded in daily life (Berger and Luckmann 1967, 66). Further, for a person to claim a position in the group he or she has a responsibility to master and use a particular technology. We are coerced by others' expectations. As suggested by the Katz principle, we are forced to work around our friends who are not a part of the system (Weiner 2007).

The weight of reciprocity can be seen when a member of the group does not use the technology of social mediation. Another perspective on this is afforded when the system itself breaks down (Star and Bowker 2006; Wurtzel and Turner 1977). As technologies become more engrained in society, we do not pay them heed. It is only when systemic disruptions occur that we understand their role. Up to that point, however, we do not focus much attention on well-embedded social mediation technologies. Until it breaks down, the telephone system, for example, is in some ways invisible. However, it gains a glaring visibility when it breaks down and we are faced with the need to communicate through alternative systems (Wurtzel and Turner 1977).

Becoming Taken for Granted

A variety of issues come into play when we choose (or are actively encouraged) to consume a technology. As the domestication approach suggests, while there are different motivations leading to consumption, the way that we integrate the artifact afterward gives us insight into important social dynamics.

Domestication can be used to understand both standalone as well as social mediation technologies. The focus here is to understand how the latter sometimes become taken for granted. History is littered with many technologies that were used to mediate social interaction, such as optical telegraphs and DeWitt Clinton's cannon fusillade, mentioned above. These technologies did not reach critical mass, adequately adjust the social ecology, gain legitimation, or enter into our shared notions of how we should carry out sociation.

Other technological systems have become—and are—social mediation technologies. Email, calendar systems, Facebook, and a broad spectrum of social networking sites are all examples. The three that are in focus here—the clock, the car, and increasingly the mobile phone—have mutual synergies as we will see in chapter 9. They let us communicate, coordinate, and transport ourselves. They facilitate the interaction and maintenance of our intimate sphere as well as our broader social lives. In short, they facilitate group interaction. The different dimensions of the internet, including the mobile internet, are also this type of technology; and indeed, that functionality is increasingly moving onto mobile devices.

There is often an emergent nature to our use of these technologies. We start by recognizing their usefulness and eventually we come to think of them as the status quo. At some critical point in the diffusion process the technologies of social mediation gain an objectivity (Gilbert 1989). At this point, they structure social interaction. Social interaction is to some degree based on the assumption that all members of the group are users. They exist outside the will of the individual, and they coerce individuals to use them. In their final state, these technologies achieve a stable and taken-for-granted role in society. Weber's iron cage has established itself.

Seen in this way, the mobile phone and the internet are still evolving. There is a certain stability associated with the PC and the mobile phone as artifacts, but even that is being challenged as the two grow into one another. The "emergent" nature of these phenomena is important. We are seeing the transition of the traditionally PC-based internet into mobile devices such as smartphones, slates, netbooks, and tablets. Yesterday's immutable need to sit beside the landline phone and await that important call has been transformed into the need to make sure that your mobile phone is charged and that you are in an area that has coverage. This again may be replaced by making sure you are logged into your location-based app so that you can follow the movements of the gang as they go from bar to bar (Humphreys 2007; Sutko and de Souza e Silva 2011; Gordon and de Souza e Silva 2011).

In all of this, there is the drive for sociation. Our desire that *others* will be available for our calls, texts, and updates is just as strong as others' desire for *us* to be available for their calls, texts, and updates. Optical telegraphs and cannon fusillades have been developed into communication channels with functionality beyond the dreams of Governor Clinton. We can easily let others know that we are on our way from Buffalo to Albany (eventually, with our two casks of water) so long as we, and they, have remembered our phones.

3 "My Idea of Heaven Is a Daily Routine": Coordination and the Development of Mechanical Timekeeping

Time and timekeeping have been a part of society for many centuries. However, with the development and refinement of mechanical timekeeping, the two are intertwined into society to the degree that we take them for granted.

The past 700 years have seen a profound transition in our sense of time. Before the development of mechanical timekeeping, there was a "natural" sense of time: activities were arranged in accordance with the rising or setting of the sun; more fine-grained activities were timed with reference to, for example, heartbeats. A mechanically based, abstract system of measuring time eventually replaced these natural systems, contributing to the increasingly well-mechanized coordination of social interaction and the development of complex logistical systems (Beniger 1986). Previous to the development of reliable mechanical clocks there were different types of water clocks, sundials, astrolabes, methods of tracking the phases of the moon, and ways of following the seasons. These were used in the coordination of agriculture and also as a way to determine the cycle of rituals and other communal events (Andrewes 2002). Mechanical timekeeping and "clock time" developed in medieval Europe. It started as a way to facilitate existing practices but soon came to be a system that framed and prefigured other activities. Those who controlled timekeeping had a hand in the control of society.

Starting with the development of simple and relatively crude devices molded to existing social practices (e.g., the cycle of prayer in monasteries), timekeeping devices developed into a broader system that today controls and coordinates social interaction. The hands of the medieval clock became a public, verifiable standard for timekeeping similar to other standards such as weight, volume, and length. Clock time became important in the coordination of commerce and the interactions of craft persons. Clocks in towers quickly became a part of the technology of profit and management. They became a way of regulating work. This trend was extended when privately owned clocks and pocket watches arrived on the scene. The person with a clock was a symbol of the industrial world, in contrast with the slower and more irregular patterns of a person living a pastoral life. Time measurement gave (and still gives) people a sense of autonomy and the ability to govern their own movements, but it also ties us to one another's expectations. Above all, the clock provides for social coordination by giving us a common metric (Glennie and Thrift 2009, 235).

I assert that time is a social mediation technology. It assists us in planning interactions with friends and family. It facilitates the coordination of work and allows us to arrange informal engagements. As with other social mediation technologies, it is not just an end in itself, but a way for us to synchronize our activities and connect and meet up with, for example, our circle of friends, family, and colleagues.

Mechanical timekeeping is perhaps the most deeply embedded social mediation technology (Elias 1992). It has diffused throughout society. Indeed, it is challenging to find any remote outpost or any social interaction where time does not play a role. Although we may have different ways of observing punctuality, the twenty-four-hour system of clock time[1] is nearly universal (Levine 1998).

We expect one another to have mastered timekeeping. According to Berger and Luckmann, the temporal regularity of society imposes "prearranged sequences" on individuals. They also say that we have a nearly "instinctive" need to reorient ourselves in those cases when we somehow lose track of time (Berger and Luckmann 1967, 42). Put in the terms of Silverstone and Haddon, timekeeping has become domesticated (Silverstone, Hirsch, and Morely 1992; Silverstone and Haddon 1996). We have legitimated the clock and timekeeping in our sense of punctuality, and the clock has reformed the social landscape.

Time is so encoded into the machinery of society that we cavalierly use it, without a thought, in our everyday affairs. It allows for the coordination of all kinds of complex organization. Following Lewis Mumford, the clock synchronizes society (Mumford 1963, 14). It is impossible, for example, to run a train line, a telephone company, a factory, a hospital, or an airport without reference to time (Thompson 1967; Mumford 1963; Zerubavel 1985, 8, 16). There is the plane that has a takeoff "window" ending at 10:23, the truck delivering bakery goods that will be at the store in a half hour, and the stoplights timed for five-minute intervals. In many cases, these sequences are organized to facilitate the flow of complex systems. In addition, the timing of the factory whistle or deadline for applications can also encode power structures (Giddens 1986, 44).

Durkheim refers to the social dimension of time as "total time":

> The concrete duration that I sense flowing in me and with me could not approach the idea of total time, for my sense of duration expresses merely the rhythm of my individual life; the idea of total time, on the other hand, corresponds to the rhythm of a life that does not belong to any individual in particular but is one in which everyone participates. . . . the rhythm of collective life dominates and encompasses the varied rhythms of all the constituent lives that produce it. And so the time that expresses it and dominates and encompasses all particular durations. This is total time. (Durkheim 1995, 337–338)

This social sense of clock time has become a given. We learn how to tell time as children. Throughout life, we unthinkingly make agreements based on the assumption that others will also use clock time. Following the Katz principle, you become a problem for me if you do not keep track of time (Katz 2008, 435).

Thus, not just time but timekeeping is a technology of social mediation. The social imperative for timekeeping exists before us and it exists outside of us. We ignore the use of timekeeping at the risk of ostracization. In the sections below, I will look at the diffusion of timekeeping devices and the use of "clock time" by large portions of society, our development of an ideology that supports timekeeping, the ways that timekeeping has changed the social ecology, and, finally, the reciprocal nature of timekeeping.

"Sonnez les matines!": The Diffusion of "Clock Time"

Before the development of the mechanical clock, there was a collection of different timekeeping devices such as water clocks, astrolabes, and sundials. For typical people, the angle of the sun, the position of the stars, or, for measuring shorter periods, the number of pulse beats gave the individual a sense of the time. Somewhat ironically, Galileo used his pulse to measure the regularity of the pendulum. The principle of the pendulum was, in turn, applied to the regulation of clocks, and now doctors use clocks to take our pulse.

Before the development of mechanical timekeeping, we did not have the ability to refer to a personal timekeeping device to coordinate everyday life. However, starting in the late 1200s, timekeeping became increasingly intertwined in everyday life. Excluding the use of water clocks, hourglasses, and the like, mechanical clocks first appeared on the scene in the thirteenth century. The motivation for the development of mechanical timekeeping devices is often linked to the Benedictine need for a reliable way to mark the cycle of prayer (Zerubavel 1985, 31). The Benedictine order had a prescribed schedule that outlined when they should work (from the first to the fourth hour), read (from the fourth to the sixth hour), rest (after the sixth hour), and so on. An ideological apparatus demanded the partitioning of sacred and profane time (Durkheim 1995, 313). According to the Order of St. Benedict, which was codified several centuries before the mechanical clock: "In the winter time, that is from the Calends of November until Easter, the sisters shall rise at *what is calculated to be the eighth hour of the night*, so that they may sleep somewhat longer than half the night and rise with their rest completed" (emphasis added). Obviously, when you do not have a machine to calculate the eighth hour of the night, this is complicated. There is, for example, a question as to when the night starts. Is it when the sun disappears below the horizon or when it is no longer twilight? How is it possible to measure this when there is cloud cover? If the night starts at sunset, it is a moving target that changes through the seasons, particularly as you move farther away from the equator. Indeed, in some climes the sun never comes up during the winter—not that the Benedictines were particularly active in places like Hammerfest or Berlevåg. In addition, before a reliable mechanical system was in place, the Benedictines needed someone to stay up and ring the bell at the appropriate time (however that was calculated). Given these complications, many have claimed that the need to calculate things such as "the eighth hour of the night" motivated the development of modern timekeeping.[2]

The earliest mechanical timekeeping devices perhaps took their inspiration from geared astrolabes (Dohrn-van Rossum 1992, 79).[3] The astrolabe is a map of the sky that traces the path of stars, planets, and the sun. In a similar way, the tracking of the sun (or, more correctly, the spinning of the earth in reference to the sun), is the job of the clock.

The mechanical clocks were designed to ring a bell at different points in the day to mark the time of prayer (Andrews 2002). An interesting if possibly oblique reference to this is seen in the French children's song "Frère Jacques":

Frère Jacques, Frère Jacques,
Dormez-vous? Dormez-vous?
Sonnez les matines! Sonnez les matines!
Ding, dang, dong. Ding, dang, dong.

A loose and abbreviated translation of the song is something like "Brother John, are you sleeping? The morning bells are ringing. Ding, dong, ding." It is the reference to *matines* ("matins" in English) in the third line of the

French version that is of interest. *Matines* is not simply morning; rather, it is the first point on the canonical cycle of prayer. Writing on fourteenth-century France, Barbara Touchman notes that *matines* was "around midnight" (Tuchman 1978, 54). In other references, *matines* is what we would call 2:00 a.m., that is, eight hours after a 6:00 p.m. sunset. In yet other references, *matines* is at sunrise. Regardless of when it was customary to recognize *matines*, it is hard to understand how someone could know that it is 2:00 a.m. or "around midnight" without a timing device.

It is not clear from the song's lyrics, or from the scholarship on the song, whether it is a mechanical device ringing the bells or if Brother Frank is up in the belfry manually ringing the bells. It has been asserted by Irvine Loudon (2002), and contested by others (Ganem and Carson 1999), that the song was written in the 1600s, giving some sense that the bells were those of a mechanical clock.

In any case, the verse points to the need for time-based signaling. This was the motivation for the mechanical clock. Assuming that the song is about one of the early mechanical clocks, it was likely very different from modern clocks as we know them. Indeed, such clocks had no face with arms, only a system to ring a bell at specific points during the day. The idea of a clock face with hours marked on it came only later. This is because the cycle of prayer—the separation of sacred and profane time—was associated not with numbered hours, but rather with the cycle of the sun. The numbering of hours associated with mechanical timekeeping was placed on top of a preexisting cycle.

The Obscure Provenance of Monastic Timekeeping Devices

Little documentation exists of exactly when the first mechanical devices appeared. References start to appear in the late 1200s but they are often obscure, and it is not clear whether the writers are referring to a mechanical clock or to some type of quasi-manual "horological" (Dohrn-van Rossum 1992, 48). This is not odd, as there was no universally recognized name for a clock at that point. With an almost audible sigh, Dohrn-van Rossum writes that "the various developmental steps of the monastic alarm devices . . . remain obscure" (1992, 60).

Regardless of when it finally arrived during the late Middle Ages, the mechanical clock did not automatically change the sense of time and periodization. However, it did start the process. Along the way it was necessary for the mechanical clock to establish itself as a reliable and useful way to measure time. Only then were institutional practices and everyday uses mapped into the cycle of clock time. Eventually clocks gained a visual dimension, usually in the form of a face with a single arm marking some form of hours. This was, however, problematic, because two approaches to dividing the day existed. During the medieval period, the norm was often to divide the day and the nighttime into two twelve-hour units. That is, the time from sunrise to sunset was twelve units, as was the time from sunset to sunrise. The length of the hours, however, was flexible. They were adjusted during the year so that there were always twelve hours of daylight and twelve hours of darkness. The alternative—which we have since adopted—was to divide the whole daily cycle into twenty-four equal units.[4] In equatorial regions, the former approach is reasonable, since the period when the sun is up varies only slightly during the different seasons. However, the further north or south we go, the longer the sun is up in the summer and the opposite during the winter.

The mechanical clock resolved the issue in favor of twenty-four equal-length hours, simply because it is far more difficult to build a mechanical device that could adjust the length of the hours depending on the cycle of the seasons (Zerubavel 1985, 37; Andrewes 2002).[5] Indeed, the early clocks were doing well to track the cycle of the day. These clocks could vary by tens-of-minutes per day and often had a sundial installed nearby to correct the clock as needed. As mechanical clocks came onto the scene, the use of the unequal hour system was slowly replaced by equal hours. In words that are evocative of Weber's iron cage, this is what Eviatar Zerubavel called the rigidification of time (Zerubavel 1985, 39).

The definitional issues with development of clock time, however, did not stop with the length of the hours. Another question was when to start the day. The current system of counting twelve hours twice starting at midnight was not a given. In some cases, the system used was the familiar twelve-hour cycle starting at the highest point of the sun, that is, noon. However, some systems had cycles starting with the rising of the sun. Since sunrise varies throughout the year, the anchoring of equal hours starting with sunrise could mean that the highest point of the sun (our noon) could vary at some latitudes from four o'clock to eight o'clock. The practice of counting the hours also saw many alternatives. In some cases people counted from one to twenty-four. In others, people counted from one to four, six times, one to six four times, and many other variations. There was also the question of how to alert people to the passage of time. Should it be audible (ringing only), visual (a dial), or both?[6] The now-so-familiar clock face was late in coming. To be sure, the first clock dial is reported only in 1521, perhaps as much as 200 years after the development of the first mechanical clocks.

The introduction of a clock face meant that clockmakers needed to decide what part of the dial moves (the pointer or the dial plate); where the "0" point for the dial should be located (top, bottom, side); how many pointers are needed (hour, minute, second, day, week, tide, moon, etc.); and whether the hour and minute dials should be imposed on one another or separate. In addition, should they include windows showing the phase of the moon, as was the case with the geared astrolabes of the thirteenth century? Even the seemingly self-evident issue of which direction the arms should rotate was open to discussion (Glennie and Thrift 2009, 139).[7] Both literally and figuratively, there are a lot of moving parts here. Until these definitional issues were settled, no single common way of "telling time" was included in the expected stock of knowledge (Berger and Luckmann 1967).

Eventually people converged on the current division of two twelve-hour cycles starting at midnight and a twelve-hour dial with a sixty-minute dial imposed on that. In addition, the most common arrangement of hands includes an hour, a minute, and a second hand, though this varies. This setup has been standard with the exception of some short-lived larks such as the ten-hour day suggested by the French during their revolution and more recently by Swatch with its idea of "Swatch Internet" time.

As noted, the original clocks developed by the Benedictine monks varied wildly. This limited their usefulness in coordination. Based on work by Galileo, the Dutch polymath Christiaan Huygens (1629–1695) developed the pendulum (and the anchor escapement) in 1656, which dramatically increased the accuracy of stationary clocks. Huygens also developed the balance spring in 1675 that opened the way for the invention of portable clocks and eventually watches. The pendulum clock ushered in what has been called the horological revolution (Landes 1983; Glennie and Thrift 2009, 35; Andrewes). Use of the pendulum meant that the clock functioned more efficiently (less winding and less maintenance), and more than anything else, it made clocks more precise. With the introduction of the pendulum, the accuracy of clocks dropped from errors in terms of hundreds of seconds per day to perhaps as few as ten (Whitrow 1988; Glennie and Thrift 2009, 251). The introduction of temperature compensation to pendulums further reduced error to as little as one second per day. This is not to suggest that all the clocks in a particular town or village were synchronized, only that they were each more precise in their own way.[8]

"Knowing the Time": The Growing Use of Timekeeping

Beyond the technical development of the clock, a major part of the story is how people came to use time and timekeeping. When, for example, were there enough clocks available so that people had the perception that there was a critical mass of clock users? E. P. Thompson (1967) claimed that this was a consequence of the industrial revolution; others have traced this perception further back in history. In the late fourteenth century, people began to gain access to timekeeping with the rise of public clocks. Gerhard Dohrn-van Rossum documents a Klondike-like rush to build secular clocks and clock towers in the period between 1360 and 1420. He notes that by 1410 approximately 25 percent of all cities in Europe with a population of over 10,000 had a public clock (Dohrn-van Rossum 1992, 165). In some cases, clocks were paid for with a common subscription.[9] In other cases, a monarch or rich benefactor supported the effort.

A critical point in the social use of time seems to be in the late fourteenth and early fifteenth century, when timekeeping was normalized and brought into everyday life. By this point, people could expect others to have a sense of time and timekeeping. By the late seventeenth century, timekeeping had become a well-accepted part of society. It is possible, for example, to see how thoroughly integrated timekeeping was by looking at situations when it failed. Paul Glennie and Nigel Thrift describe the quick reallocation of local budgets in the 1600s whenever the local clock in the clock tower broke down. They note the implementation of makeshift alternatives that included borrowing a clock, presumably from a neighboring parish. Other stopgap solutions included manually adjusting the hands based on the readings of hourglasses and paying people to strike the bell when the striking mechanism of the clock was not in order (Glennie and Thrift 2009, 143). This is not to say that all social synchronization was stranded when the local clock failed (Zerubavel 1985, 21). Coordination within small groups that was highly routinized likely continued as normal. However, in those situations where there was a need to organize larger groups of individuals engaged in more complex tasks, the loss of a standard system of timekeeping was more dramatic. The fact that towns were willing to mobilize resources when the clock failed speaks to the way that timekeeping had become embedded into the structure of mundane life.[10]

The diffusion of personal timepieces in society replaced the reliance on a single, central clock by making the timekeeping system more redundant. Based on the developments of Huygens in the 1650s, the clock became more portable. Soon it had been diffused into society to the degree that it was seen as essential. It had moved from being a way to mark the prayer cycle to being a central coordinating tool. By this time the clock was an assumed part of the public sphere.

> Although the actual workings of clocks remained unfamiliar [after the horological revolution], even mysterious or magical to many people, the notion of linking one's conduct to certain hours was something so common so as to have attracted little comment. This has often been interpreted as indicating that most people could not comprehend clock time, but to us the moral is different. It was precisely *because* clocks were a familiar part of many everyday landscapes that they did not need to be explained. (Glennie and Thrift 2009, 176)

Indeed, according to Glennie and Thrift, understanding of clock time was widespread, and derived from the tradition of tower-based clocks. The newly portable clocks extended the range of timekeeping beyond the audible sphere of the tower bells or the immediate view of the clocktower.

> That clock times were already involved in many familiar practices for much of the population helps to explain the otherwise puzzlingly rapid diffusion clocks and watches between 1660 and 1730. They were newly cheap(er) items, or newly reliable items, available to the population able to deal, and in many places long-used to dealing with clock time . . . large parts of the population already have relatively sophisticated understandings of everyday clock time and what this could entail. (Ibid., 177)

By the 1730s, clocks were found not only in church towers and other outdoor locations but also in pubs, churches, and inns. Time and timekeeping had become part of the coordination regime. Between 1600 and the mid-1700s, timekeeping became increasingly central to the planning of everyday interactions. People were using time to coordinate their lives. Put into the terms of social mediation technologies, timekeeping was used as the basis of reciprocal expectations. Timekeeping had achieved a critical mass. It was (perhaps coercively) governing peoples' mutual expectations.

Timekeeping also moved into the domestic sphere. According to an analysis of probate inventories associated with wills and an examination of clockmaking apprenticeships in the UK, Glennie and Thrift (2009, 164) have found that clocks started to appear in homes in the late 1600s and that there was an increase in access from the middle of the 1700s until the start of the 1800s. This reflects indirectly the growth of time-based regulation, and it undoubtedly also points to the symbolic value of owning a clock. Interestingly, people in the UK of all social levels owned clocks.

> Overall, in early Georgian Bristol (the mid 18th century), then, numerous dimensions of everyday life were riddled with clock times, with mutually reinforcing impulses from work, from transport, from leisure, and from everyday sociality. Dense networks of various kinds of timepieces, and other devices, made "knowing the time" a routine matter for people in most parts of the city. The making of distinctions in hours and minutes was pervasive, if largely taken for granted. There are few indications that telling the time was an unfamiliar practice, or that people might be

disadvantaged by an inability to do so—that too was taken for granted. (Glennie and Thrift 2009, 122)

This is not to say that the use of time and timekeeping was the same in England as in Spain, Germany, India, or Mexico. In each country, and indeed in each city, local practices dictated how to count the hours and the arrangement of the clock dials; these practices were slowly being merged into the single, more abstract system of timekeeping we know today. Sometimes this went easily; in other cases, it was accompanied by kicking and screaming. Regardless, by the beginning of the modern period in Europe, clock time was well entrenched.

The trend toward the ever-increasing personalization of timekeeping led eventually to pocket watches. These in turn permitted new forms of coordination. Pocket watches were largely handmade and thus very expensive. The push to develop chronometers for navigation in the mid-1700s made clocks more portable, precise, and robust (Sobel 1996). Factory-produced pocket watches were first seen on a broad scale immediately before the American Civil War (1861–1865) (Sandvig and Sawhney 2004), as it was only then that manufacturing processes were able to operate at the needed level of precision.[11]

Although the reduction in size made the mass adoption of clocks possible, mass production actually brought it about. In fact, the mass production of clocks spawned a whole array of industrial techniques. It gave birth to batch production, interchangeable parts, the use of sub-trades, tooth- and wheel-cutting machines, and many advances in metallurgy and metal working. (Sandvig and Sawhney 2004, 361–362)

The Waltham Watch Company was able to develop the process of machining all the pieces for a watch in the period leading up to the Civil War. This meant that the cost of watches fell and, as a result, many young men going off to fight the war owned a watch. This development allowed for a more refined method of planning and coordination on the battlefield. The personalization of timekeeping devices allowed generals to coordinate large groups of soldiers.[12] The cannons at Spotsylvania or Vicksburg would shoot until 10:00, the infantry would move into place by 10:30, the right flank would attack at 11:00, and so on (Carosso 1949, 170; Andrewes 2002). Ironically, mechanical timekeeping, which had arisen as a way to separate sacred and profane time (Durkheim 1995, 313), was now being used to coordinate slaughter.

Timekeeping had long been broadly available. At some point during the diffusion of clocks, timekeeping reached a critical point. The process leading up to that transition was characterized by reliance on natural cycles or other time regulation regimes (e.g., manually rung bells, water clocks, sundials). Eventually there developed a critical mass of people who had access to clocks and understood how to tell time. This transition led to the embedding of timekeeping in society.

Helping "Citizens of Honor to Lead Orderly Lives": The Legitimation of Clock Time

As a technology diffuses and increasingly imposes its logic, the need to justify or legitimize its use arises. Following Berger and Luckmann, "the institutional world requires legitimation, that is, ways that it can be 'explained' and justified" (1967, 61). With the transition from natural time to "clock time" arose the need to develop arguments supporting the change. Timekeeping needed to be situated in the context of other broadly understood social principles.

By the late seventeenth century, after the horological revolution, there was a well-developed sense that time, and the respect of time, was associated with courtesy, punctuality, and morality. This helped to facilitate complex forms of cooperation and, in a broader sense, it was a key for understanding the universe and the role of humanity. There was the sense of punctuality in the Benedictine system. The phasing of different activities was to observe their "proper" or "appointed" status. The monks assigned positive value to regularity and punctuality (Zerubavel 1985, 35). This later developed into a sense of etiquette. The conflation of punctuality and our sense of courtesy legitimates the use of clock time.

> Fixing the temporal location of events entails a broadly conceived norm of "punctuality," which involves assigning a deviant character to the acts of being "early" or "late." Being late for work, for example, might entail some loss in one's reputation. (Zerubavel 1985, 8)

Robert Levine notes that until the development of an agreed-upon standard for timekeeping and the method with which to measure this, punctuality was a very loose concept (1998). Indeed, we had no easy way to know whether we were hopelessly early or late before there was a time standard. It was not easy to hold others accountable to precise points in time since there was no common measure. Indeed, the word "punctual" did not refer to precision in timing until after the horological revolution.[13] Once we had this concept, it did not take long for people to associate punctuality with courtesy. Dohrn-van Rossum quotes a Spanish writer from 1392 who noted that a—perhaps contrived—justification for early city clocks was to allow "citizens of honor to live orderly lives and to call sleepers and idlers to virtuous works" (Dohrn-van Rossum 1992, 137). In a similar vein, to citizens of early modern England punctuality was seen as a social grace.

There was, and often still is, a sense of moral order that arises from being prompt (Glennie and Thrift 2009, 189). In some cases, this was taken to the extreme. Moving to the metaphysical level, not only did individuals agree about when they should carry out different activities with other people, but on occasion people reported that angels used clock time to determine meeting times. For example, Bartholomew Hickman reported hearing the voice of an angel calling him into conference: "Tomorrow half an hour after 9 of the clock, give your attendance to know the Lord's pleasure" (original in Fenton 1998; cited in Glennie and Thrift 2009, 212) It would certainly be a breach of manners to abuse the punctuality of such an invitation. It is also surprising to note that divine authorities were using clock time. In some cases, the increased precision afforded by clocks was a phenomenon in search of a use, and the precision afforded by mechanical clocks was applied in interesting directions. In the case of astrology—a system of thought that had received new vigor with the development of the printing press (Eisenstein 1979, 437)—births were recorded to the minute in order to facilitate later examination of how the arrangement of planets could be interpreted. Some people took this impulse to its logical conclusion and even noted the exact time of sexual intercourse, presumably in order to understand how the stars would affect conception (Glennie and Thrift 2009, 202). Thus, the precision afforded by timekeeping was applied to an increasing number of sometimes more and sometimes less fanciful applications. In these practices, we see the use of precision in the investigation of human nature and its application to an individual's presumed direction in life. There was an assertion of a link between these elements. Regardless of the real efficacy of the link, we can see people asserting the legitimacy of clock time as an instrument in the analysis of the human condition.

By the twentieth century, with the diffusion of less expensive factory-produced watches, there was an increase in the demand for punctuality, as well as a corollary sense that being late was a sign of social inferiority and incompetence (Levine 1998). While there is some latitude to the sentiment, we still accept the idea that being on time is a matter of courtesy. Exactly how punctuality is arranged in concrete social situations can lead to problems. Quoting people interviewed in the contemporary UK, Jenny Shaw provides us with insight into people's sense of timeliness and punctuality:

> I've tended to make an "art" of being punctual. I have little time for those people who keep me waiting. My time is valuable too, and I consider those who waste it damned selfish. . . . Put another way, in the Royal Navy or the Merchant Marine if a man missed a ship because the bus or train was late, the car had broke down or the wife had a baby, that ship was a man short. . . . Just picture the scene and imagine what the captain would say if the executive officer said to him "Sorry, Sir. We can't fire the missiles or fly off the helicopter; half the crew missed the last bus." I can assure you the reply would be unprintable and the situation unthinkable. (Quoted in Shaw 1994, 90)

Another of her informants says:

> If people are not on time I get terribly irritable and as unhelpful as I can. Not too good in this country where hardly anybody bothers to be *on time*. Most people consider "on time" to mean 15–20 late . . . My idea of heaven is a daily routine. . . . I am pathological about time keeping; never forgive anyone for being late, even by a few seconds. (Ibid.; emphasis in the original)

Shaw also quotes yet another woman who says, "My husband is a lunatic over time keeping" (1994, 86). This woman goes on to say that unpunctuality and untidiness are signals of those who shirk social responsibility. She notes that we can be untidy without necessarily trespassing on others' sensibilities, but unpunctuality is a social failure (Shaw 1994, 93–94).[14] Punctuality becomes an arena in which we work out issues of power.

While punctuality is seen as an important characteristic, it is also clear that arguments have been made against the use of clock time. People have said that the rigid observation of clock time abuses the needs of individuals. The idea of natural time and discussions regarding the tyranny of the clock are common themes (Thompson 1967; Eriksen 2001; Mumford 1963). In the tension between clock-driven punctuality and the idea of natural time, the forces of punctuality seem to have taken the day. Nevertheless, there remains a certain tolerance for lack of precision that can vary from culture to culture (Levine 1998; Reis 1980), between different groups within a society (Zerubavel 1985, 67), or across genders (Kristeva 1981).

Thus, a social legitimation process has arisen along with the development of the mechanical clock. We see this in questions of manners and the use of punctuality, and we see it in the aesthetics and the sense of morality that is associated with clocks as artifacts. Such attitudes toward time and punctuality legitimate the use of clock time. The fact that we have these rules underscores the social nature of time as a social construction.

The legitimation of mechanical timekeeping is also seen in the effort put into building clocks and timekeeping instruments. The very design and decoration of clocks is a type of legitimation. This can be seen in the exquisite handwork of public clocks as well as smaller timepieces in the home. Even today, we display expensive or vintage grandfather clocks and mantelpiece clocks in the "public" portions of the home, in the hallway or the parlor, and we wear expensive and bejeweled watches. These devices give evidence of the town's, the family's, or the individual's sense of aesthetic.

Some large tower clocks provide a further elaboration of morality. For example, some public clocks, such as Orloj in Prague, utilize automated biblical figures to illustrate different moral points. In this case, the striking of the clock unleashes a small morality play involving various devils and saints. In addition to telling time, Orloj contains a series of sculptures representing different themes including threats to goodness.[15] When the eyes of the populace are checking the time, they also get a small dose of religion. The crafting and effort put into this clock speak to how timekeeping was legitimated in medieval Prague.

Finally, some mechanical clocks assert the special position of humanity among God's creations. This is seen in devices that replicated the functioning of the universe (Friedman 1984). This was perhaps taken to its logical conclusion by the Italian Giovanni de' Dondi, who took sixteen years to build his *astrarium*, a device that represented the movements of the seven known planets on each of its seven dials, albeit using the Ptolemaic, not the Copernican, universe. Indeed, using the Ptolemaic perspective of having the earth and not the sun at the center of the universe makes the construction of the device all the more complex (Bronowski 1973). In addition to the planets, it had a twenty-four-hour dial and a calendar showing the date along with the different fixed and movable feasts of the church (Bedini and Maddison 1966). Finally, it also showed the position of the zodiac and moon.[16] This masterful construction played into the sense that so long as people applied intellect and skill to the development of machinery, they, among all creatures, could master nature and were most Godlike. The idea was that by carefully tracing the movement of the heavens and measuring time, humanity was somehow touching the hem of divinity. Timekeeping was not only a practical affair; it was an affirmation of our special role in the universe. It was perhaps the ultimate the legitimation of timekeeping.

The Temporal Regularity of Modern Society: The Social Ecology of Timekeeping

Coordination of society with the use of mechanical timekeeping went from the guiding of prayer cycles to being an indispensable part of society. It replaced the logic of natural cycles as well as the customs of independent institutions—each having its own timing system—with a common metric used to regulate and coordinate ever more complex interactions. In this way timekeeping changed the existing social ecology. According to Zerubavel:

> we are able to treat our time units as standard quantities of duration mainly because we can measure them against a timepiece that is paced at a uniform rate. It is thus the clock, whose introduction to the West cannot be separated from the evolution of the schedule there, that allows the particular notion of temporality which has become so characteristic of Western civilization. It is *clock time* that is at the basis of the modern Western notion of duration and that allows the durational rigidity that is so typical of modern life. (Zerubavel 1985, 61)

After they became common, mechanical clocks replaced other systems in a wide variety of tasks. They were used to time the opening and closing of markets, the timing of council meetings, and the coordination of complex construction projects; they also served as the basis for measuring the time used for work and thus were the basis of payment. To an increasing degree, it was not just a gentle prompt that one or another ritual or activity was about to begin. Rather, clock time became the mechanism with which to coordinate increasingly complex processes (Dohrn-van Rossum 1992, 323; Zerubavel 1985). Clock time overtook the niche for coordination by out-competing sundials, water clocks, manually rung bells, and other timekeeping systems.

Following Glennie and Thrift, clock time is a set of "embodied practices" that imposes its own logic on society and is an enmeshed part of the ongoing practices of social coordination (Glennie and Thrift 2009, 72). As clocks and watches became increasingly accurate, they replaced (or perhaps out-competed) other forms of timekeeping. As this process continued, social interaction was increasingly adjusted and controlled by the clock.

Mechanical timekeeping and the system of abstract time units also replaced the more idiosyncratic temporal regulation of local institutions. Most notably, mechanical clocks replaced the extensive system of manually rung bells that notified citizens of different events in the city. Prior to the rise of mechanical clocks, it was common for European cities to use a system of bells—or what Dohrn-van Rossum (1992, 203) calls an "acoustic environment"—to alert people to various events. Manually rung bells were used to mark the cycle of prayer, the opening and closing of gates, starting of church services, opening and closing of markets, changing of guards, and the beginning of council meetings.[17]

Each city, and indeed many churches, town halls, and public buildings, had a range of bells. A particular building might have some bells with a low tone, some in the middle range, and some that were high pitched. Each bell was used in a particular situation. In many cases the individual bells had names and a known legacy. Before it gained notoriety for its jaunty angle, the Leaning Tower of Pisa was planned to be a bell tower with seven different bells, each having a name and a specific purpose. The bells were used individually and collectively to communicate different messages. For example, one bell, named *Pasquereccia,* was used to mark Easter. In its earlier location on the Tower of Justice, the bell had been used to announce capital executions of criminals and traitors. A city such as Florence had more than eighty different bells, each of which was used to notify the public or mark an event. Further, to the confoundment of those who were not familiar with the local practice, each city had its own special regime. Mechanical timekeeping challenged this complex—and in many cases unwieldy—system with a unified system of signaling. Where bells used acoustic signals targeted at specific audiences, timekeeping provided a common standard for all.[18]

With the adoption of a town clock, or perhaps a nearby monastery's clock, some people began to use clock time to keep track of the day. People started to rouse when the clock struck for matins. They routinely stopped for lunch when the clock struck noon, and so on. If, for example, a worker was to receive a certain payment for a half-day's work, there may have been reference to the striking of the town clock to determine the correct period of work. Eventually, enough people referred to the local clock that it became a commonly understood reference point. When it achieved this status, it moved from being an oddity that only some people used to being the status quo. This is not to say that it was accurate or that the workings of the clock were understood, or even that everybody could use it as a common point of reference. It is clear that there would be, and indeed still are, discussions as to the need to be strict in observing time or the accuracy of a particular timepiece (Levine 1998). But nevertheless, when the clock time became the accepted state of affairs, the clock had moved from being a useful curiosity to being an assumed part of life. Clock time had asserted itself as a primary organizing principle for a particular locality, and alternative systems were put into a secondary position. Clocks had changed the social environment.

By the seventeenth century, mechanical timekeeping replaced the chaotic system of signaling bells with its own logic. Where there had been independent signaling systems to coordinate and sequence various activities, mechanical timekeeping replaced these with a single unified system. This meant that people had to adopt the more abstract system of clock time. The advantage was that this transition allowed for more complex coordination. Both in urban and in more rural settings, mechanical clocks were being used to determine when different tasks and activities were to take place (Dohrn-van Rossum 1992, 237). Rather than simply marking the opening and closing of a market, mechanical timekeeping allowed the coordination and sequencing of skilled workers around a central task, the cycling of livestock in their grazing, and the calculation of wages and fees (Glennie and Thrift 2009, 219). There was also a growing rationalization (what Zerubavel [1985, 50] calls "durational rigidification" and what Weber would call the iron cage) of larger systems that eventually led to the factory method of production (Glennie and Thrift 2009, 223–233). In short, abstract clock time was being imposed on the preexisting means of social coordination that had, in many cases, involved the increasingly unmanageable system of bells wherein every institution had its own special standard and cadence. Timekeeping and mechanical clocks cut through the need for this multitude of conflicting schemes and substituted a single albeit somewhat abstract system that could be tailored to the needs of the various institutions and actors.[19]

Dohrn-van Rossum (1992) and Glennie and Thrift (2009) assert that the logic of timekeeping became a central form of social coordination in the late fourteenth century. E. P. Thompson (1967) and Lewis Mumford (1963) asserted that the transition to mechanical timekeeping is one of the keys for understanding the industrial revolution and the imposition of commercial rationality.[20] Thompson posits that mechanical timekeeping communicated discipline to the workforce and was a form of power that was (and is) exercised over the workers. The factory—or for that matter urban life—demanded a level of coordination and social organization that made timekeeping more central (Urry 2007, 192). Thompson's analysis speaks to the convergence of the mechanical basis for timekeeping with new forms of production and the extensive use of time-based wage labor. These time regimes engendered new understandings of coordination that were also adopted in other institutions such as the school system. Thompson dates this transition from the late eighteenth century, when people were acquiring private access to more precise time devices (Thompson 1967; see also Giddens 1986, 136; Mumford 1963).[21]

Beniger (1986) has discussed how the tighter temporal integration of various logistical systems resulted in the need for different forms of control. The speed of both the steam engine and the manufacturing processes meant that the production system could not be managed without the development of advanced controls. In Beniger's estimation, these included the standardization, bureaucracy, and the mechanical processing of information that eventually led to information and communication technologies.

Mumford makes the same point. In his direct way of stating things, he wrote, "the clock is not merely a means of keeping track of the hours, but of synchronizing the actions of men" (Mumford 1963, 14). He traces the impact of this on the lives of individuals:

> The first characteristic of modern machine civilization is its temporal regularity. From the moment of waking, the rhythm of the day is punctuated by the clock. Irrespective of strain of fatigue, despite reluctance or apathy, the household rises close to its set hour. Tardiness in rising is penalized by extra haste in eating breakfast or walking to catch the train: in the long run, it may even mean the loss of a job or advancement in business. Breakfast, lunch, dinner occur at regular hours and are of definitely limited duration: a million people perform these functions within a very narrow band of time, and only minor provisions are made for those who would have food outside this regular schedule. (Mumford 1963, 269)

Not only was the individual subject to the dictates of the clock, he or she was also subject to the dictates of the person who controlled the clock. Indeed, Sandvig and Sawhney (2004) describe the manipulation of factory clocks by unscrupulous managers in order to adjust working hours and the need to pay wages.

The way that timekeeping had asserted its logic on the social ecology can perhaps best be seen with the establishment of global time zones (Urry 2007, 96; Blaise 2000). With the coming of the railway, reliance on a local time system was inadequate. Up to this point, there were no standard time zones, but rather hundreds if not thousands of local definitions of time based on measuring the zenith of the sun at this or that location.[22] For example, when it is solar noon in London, that is, when the sun is at its highest point, it is still about ten minutes to noon in Bristol, 2° 35' west of London. Railroad schedules and timetables were difficult if not impossible to construct and, more to the point, it meant that bungled schedules (which were frightfully common because of the complexity of the system), could result in two trains occupying the same tracks at the same time, with unfortunate consequences.

These complexities led to the replacement of local time measurement with twenty-four time zones around the world (Blaise 2000).[23] After many years of meetings, discussions, and conferences, on November 18, 1883, the global system of time zones was established. According to Carey (1988), on this day, at what had been noon in Chicago (that is, when the sun was directly overhead) the railroad system was halted for nine minutes and thirty-two seconds. By that time, the sun was directly above the ninetieth meridian that is west of Chicago: this was the basis for time in the newly established Central Time Zone.[24] The clocks were then restarted and the entire railroad system was integrated into a series of unified time zones instead of the multitude of local time zones that had existed a few minutes before.

This change was a technical one, and it was not necessarily accepted at the local level with the same speed with which it was effected. Indeed, Carey writes that "The changeover was greeted by mass meetings, anger, and religious protests" (Carey 1988, 26). In the era of horse and pedestrian transportation, locally based time was perfectly adequate. In the words of the Lynds, speaking of a perhaps overly bucolic vision of Middletown, "few owned watches, and sun time was good enough during the day, while early and late candle lighting served to distinguish the periods at night" (Lynd and Lynd 1929, 11).The transition from local time to time zones illustrates how mechanical timekeeping had moved to the center of society. It was no longer merely a convenient aid for reminding monks of their responsibilities to the prayer cycle. Rather, it had become the method for regulating a vast railroad system. Mechanical timekeeping coordinated the movement and switching of steam trains moving at high speeds. It facilitated the scheduling and ticketing systems and the expectations of passengers. Time was important in ensuring the safety of passengers and crews. It became a part of contracts dealing with the transportation of goods and delivery of the post. In short, there was a whole system of interconnected practices referring to the metric of the clock. The rest of society followed.[25]

We still live within this legacy. In contemporary society, we need to coordinate large systems such as stock trading, airline transport, navigation, administration, health provision, and just-in-time manufacturing. In each of these and hundreds more systems, we rely on a standard time and on accurate timekeeping devices. In modern society, clocks and timing have gained an even more central coordinating role. For example, the extremely accurate atomic clocks regulate the Global Positioning System (GPS) and the interactions in the global telecommunication systems. Indeed, many clocks are nothing more than a radio-receiver set to receive the time signal from a central device. All the "clocks" that happen to be listening display time that is more regular than the spinning of the earth. With the use of this information, stock markets open and close, planes take off and land (occasionally on time), and parents deliver the local soccer team to cross-town soccer fields while guided by GPS. In each case, large systems coordinate their interactions with reference to time and timekeeping. Clock time has moved from the edges of society to being one of society's tacit elements. Although occasional sundials, church bells, hourglasses *qua* egg timers are still in use, we collectively rely on clock time.

The Regulation of Behavior: Reciprocal Expectations and Mechanical Timekeeping

As clock time became a central part of everyday life, it also became a reciprocally anticipated element in the coordination of social interactions. That is, we came to expect it of one another. In this way, the use of clock time is a technology of social mediation. We have reciprocal expectations regarding the use of time that facilitate interaction but at the same time set the conditions for sociation. The diffusion of mechanical clocks reformulated the social ecology, and we have constructed justifications for abiding by clock time. Finally, as will be discussed in this section, we have developed reciprocal assumptions regarding the use of clock time. Like Weber's iron cage, our reciprocal expectations, as they gained facticity, slowly formed into a rigid constraint on behavior. Now it is very difficult, if not impossible, to ignore timekeeping in our interactions with others. Anticipations regarding timekeeping are deeply engrained in daily life to the degree that it is taken as a given. According to Zerubavel, "Our everyday life would not have been possible had we not internalized a certain interpretative order. Obviously, we are usually unaware of this order, because we take it for granted" (Zerubavel 1985, 19; see also Glennie and Thrift 2009).

Nearly all of us share the supposition that we, and those with whom we interact, attend to time and timekeeping in broadly the same way. Norbert Elias said:

> By the use of the clock, a group of people, in a sense, transmits a message to each of its individual members. The physical device is so arranged that it can function as a transmitter of messages and thereby as a means of regulating behavior. (Elias as quoted in MacKenzie 2002, 93)

The socially constructed notion of clock time regulates behavior not only of individuals, but also of groups. It frames interaction and helps to define etiquette. It makes obvious power relations and is an integral part of coordination. Simmel recognized this; although he was speaking of urban life, his comments can easily be extended to all of society:

> The technique of urban life is generally not conceivable without all its activities and reciprocal relationships being organized and coordinated in the most punctual way into a firmly fixed framework of time which transcends all subjective elements. (Simmel 1903 [1971], 328)

To illustrate the centrality of timekeeping in our lives, consider a thought experiment: what would it be like if all the world's watches and clocks were each set to their own random time (Simmel 1971)?[26] The individual devices themselves would continue to function. Each would mark the hours, minutes, and seconds, and each would require that we wind it or change the batteries at a given interval. However, since each device would be "following its own drummer," to use the phrase of Henry David Thoreau, they would be more-or-less useless in the coordination of interaction. My 12:30 would be another's 5:36 and yet another person's 23:08. It would be impossible to synchronize meetings and events. To make the thought experiment increasingly absurd, we could imagine that each of the watches and clocks has a different periodization, one with the day divided into twenty-four hours, another dividing it into ten, and a third dividing it into seventeen hours, for example. The point is that we would be unable to organize our lives.[27] The thought experiment also indicates how we take for granted a commonly accessible and commonly metered form of time. Knowing that others have the same time reference allows for coordination. When I agree with a colleague to meet at noon, I do not need to specify a timekeeping system. I do not need to be concerned with regard to the synchronization of watches, as I can assume that he/she is using the same time metric.[28] We both accept the efficacy of the timekeeping system. To ignore it would expose us to what Durkheim calls "its coercive power" (1938, 51). This tacit reciprocal agreement regarding timekeeping shows that it is a technology of social mediation.

The Compelling and Cohesive Nature of Clock Time

Clock time has become the central method for the coordination of society. Indeed, it is seen as something approaching a law of nature or the manifestation of divine will. It is, however, a social construction:

> even though the socio-temporal order is based, to a large extent, on purely arbitrary social convention, it is never the less usually perceived by people as given, inevitable, and unalterable. (Zerubavel 1985, 42)

We think of mechanical timekeeping as something that exists outside of human agency. But this is to negate the extensive social processes that have resulted in our system of clock time. An evolution has occurred that is, from our contemporary perspective, covered with such a thick legacy so as to obscure just how these social processes took place.

We have applied this metric to the way we work, the coordination of our institutions, and the regulation of our schools. Clock time sets the constraints with regard to scheduling classes, observing store hours, production schedules, coordinating deliveries, establishing time tables for trains, buses and airplanes, and governing our expectations of one another's punctuality. We use clock time to work out the particular logistics of each errand. When does the truckload of bread need to arrive at the grocery store? What time will school start the first day after a holiday? When will the night shift start in the intensive ward during the second week of June? What time will the knitting group meet? When is the pilot of flight 342 to Albuquerque allowed to start taxiing out to runway 3? When should the anesthesiologist start administering medicine to Mindie Clayton in operating room 3? What time does the news start on channel 6, and when is soccer practice? The metric of time facilitates these and an endless number of other activities.

Each one of these activities has its logistics. We can work them out in several different ways. The first is to simply follow the weight of routine (e.g., the knitting group has always met on Thursdays at 7:30 p.m. It has been that way since 1964 when it started).

The second approach is to work out and publish a particular schedule. This might be, for example, the staffing of employees for the late shift on the intensive ward or the time and location of soccer practice. This type of scheduling has been facilitated by the internet, and now internet by smartphones. In its role as a remote-access repository of easily updatable information, the internet is a perfect tool for scheduling quasi-regular activities. A colleague on the intensive ward might be sick, or the timing of soccer practice might have to change as other events arise. The internet's ability to provide a centralized and authoritative source of information is an important function of the system. The ability to access it via an app on a smartphone, wherever we might be, further facilitates the diffusion of information.

There are untold numbers of hobby and volunteer organizations that use the internet to coordinate their schedules in this way. Clubs such as the Pair-A-Dice Cruiser car club (meetings held every second Tuesday of the month at 7:00 p.m., usually at the Choo Choo Bar and Grill in Loretto, Minnesota),[29] Team Sun Things, a synchronized skating club in Cape Town, South Africa,[30] and the Forsyth Extension Sewing and Needle Arts Club[31] in Forsyth County, North Carolina, are examples. In addition, the logistics of innumerable businesses, public transportation services, and organizations rely on schedules posted on the internet (Palen 1998).

Finally, a third approach to scheduling is to work it out in real time. As we will see in chapter 7, it is in this context the mobile phone especially allows for microcoordination (Ling 2004). In all of these forms of scheduling, time is used in one way or another.

When thinking broadly about technologies of social mediation—not just mechanical timekeeping—we find that some are quite standardized and stable, whereas others are much less so. Emergent technologies are often somewhat pliable. They are perhaps more similar to Weber's "light cloak" that has not yet formed into the iron cage. This is, however, not as true of time and timekeeping. While there might be some mutability associated with time and the development of microcoordination, over the past 700 years, time and timekeeping have gained a central and deeply ingrained social role. We are not at liberty to ignore timekeeping.

Starting with the development of simple and relatively crude clocks that were molded to existing practices (e.g., the cycle of sacred time monasteries), timekeeping devices developed into a broader system that now controls and coordinates nearly all social interaction. Such devices showed their usefulness and became increasingly integral to the functioning of society. The pendulum, improved mastery of metallurgy, and standardized manufacturing made timekeeping more reliable, robust, and reasonably priced. This entailed, in turn, even greater access to timekeeping devices in both public and private life.

Yet another layer is imposed on this history, namely, our legitimation—and in some cases vilification—of timekeeping. Timekeeping is tied to issues of morality, courtesy, power, and stress. At a more abstract level, time and timekeeping have been seen as a metaphor for the ingenuity of humanity. The ability to build machines that mirror the movement of the heavens and mark the regularity of the earth's rotation was seen as being godlike. We are fond of thinking that this accomplishment reflects on humanity's mastery of nature.

Imposed on the diffusion of this technology is another history, namely, the way in which we have expanded the use of mechanical timekeeping's niche so as to influence and control other activities. Timekeeping shifted from being a system to remind people of prayer cycles to being a system that set the conditions for other activities (the coordination of the railways, factories, schools, commerce, etc.). It became the premise upon which other complex logistical systems were founded. This is obviously not to say that timekeeping in these systems respects the needs of individuals; in many cases, it does not. But it is difficult to imagine coordinating contemporary society without the help of time and timekeeping.

We are generally not free to ignore time and timekeeping. We do so at the peril of offending friends, family, bosses, and colleagues. We know that their expectations will be frustrated if we disregard time and timekeeping, and our sense of these expectations keeps us in line. Using Durkheim's test, timekeeping has become "a compelling and coercive power by virtue of which, whether he wishes it or not, they impose themselves" (Durkheim 1938, 51).

4 "Four-Wheeled Bugs with Detachable Brains": The Constraining Freedom of the Automobile

I now turn to the automobile and its symbiotic partner, the suburbs. The automobile is not as universal a technology as clock time. It is possible to live a good life without a car. Indeed, millions of people do. In places with adequate public transportation, where there are alternatives such as bicycling, or in those locations where supplies are within walking distance, the car is not essential. This does not diminish the fact that, for many people, the automobile is an essential part of life. Major portions of our cities are given over to the car. Kenneth Boulding once said,

> A somewhat casual observer from outer space might well deduce that the course of evolution on this planet had produced a species of large four-wheeled bugs with detachable brains; particular animals which rested when they sent their brains away from them but performed in rather predictable manners when their brains were recalled. (Boulding 1980, 148)

There has been little to contradict what Boulding's "casual observer" might have thought. In many ways, cars structure our lives around their demands.

The automobile is not necessarily a technology that increases social interaction. Indeed, given the isolating effects of the suburbs, the sum effect may be just the opposite (Bakardijeva 2003; Barber 2006; Hayden 2003; Putnam 2000). It may give us something similar to what Raymond Williams (1974, 18) calls "mobile privatization."[1] The private automobile and the system of urban settlement it spawned have constrained and formed sociation. This said, it is also true that we mediate sociation through private transportation (Urry 2000, 190). Like the mobile phone, the clock, and the internet, we use the automobile and its accompanying system of personal transportation to effect and facilitate social interaction.

For many, the automobile is a requirement. If you live in a city with a subway or a well-developed bus system (say, New York, London, Paris, or Tokyo), you do not need a car. If you live in a location that is bikeable (e.g.,

Copenhagen or Amsterdam), you do not need a car. However, those living in the vast suburbs of North America, Europe, and Australia have few other options (Song et al. 2010). If we want to socialize, if we want to commute to work, if we want to transport our children to their after-school activities, if we want to shop for food and other goods, a car is a necessity. Thus, in addition to thinking of the automobile as a type of artifact, it is useful to think of it in terms of a sociotechnical complex that includes roads, settlement patterns, marketing (think shopping malls), and a policy-decision apparatus that encourages this particular form of mobility.[2]

The story of the automobile is in reality an international one. Major contributions have been made by Germany, France, Japan, Sweden, Russia, Italy, China, Korea, the UK, and a host of other countries. There is no denying this. Germany first perfected the gasoline engine, and Russia produced an endless number of Ladas. France developed the nascent automobile into one that was commercially viable; indeed, their legacy lives on in a variety of words that we associate with cars, including chassis, chauffer, carburetor, coupe, limousine, and indeed the word automobile itself (Bryson 1994, 198; Eckermann 2001). China is poised to become the world's largest automobile market, followed perhaps by India. That said, until the beginning of the twenty-first century, the world's auto industry has been dominated by the United States. The two World Wars severely hindered the manufacture of cars in Europe and Japan. Prior to the First World War, 80 percent of automobiles on the road were produced in the United States or Canada (Flink 2001, 25). In addition, at critical points of development in the 1920s and 1950s, US cities were malleable to the influences of the automobile. The patterns of streets, the potential for highways, and the area available for suburban development were all conducive to the technology. Public policy was also oriented toward the private automobile and single-family home ownership. These influences meant that by the 1960s, many people in the United States were dependent on automobile-based mobility. Indeed, a sense of inevitability accompanied the development of the suburbs (Urry 2007, 114).[3] The result of this is that the automobile has become an unavoidable part of daily life for people living in the suburbs (O'Connor and Maher 1982). Regardless of whether these people want a car or not, they need one (Hayden 2003, 157).

The car has changed and indeed constrained sociation. We need it in order to participate in daily life. Many people do not have reasonable alternatives; when we suddenly do not have a car, we may have to default on our social obligations. If our car fails, our social radius is reduced and we are unable to carry out our plans. When the car breaks down, we have to work out laborious alternatives, and perhaps we become a millstone around the necks of our friends who agree to transport us.

This is not to apologize for the car. It threatens to pollute our planet beyond recognition and to kill us in car crashes (Urry 2000, 13). In spite of these problems, however, we seemingly cannot get enough of the automobile (Urry 2007, 118). Our urban environment is irretrievably designed to support automobile-based settlement. Because of this perhaps unfortunate dependency, we can see that the car and the system of institutions that surrounds it are a technology of social mediation.

The Puffing Devil and the *Orukter Amphibolos*: The Diffusion of Cars in Society

Reliance on timekeeping grew from the instillation of clocks in church towers and town halls. It was only later that clocks became items of personal consumption. Clocks in clock towers are a type of public good. They are available for all to see and to integrate into their routines. By contrast, automobiles—but not necessarily the roads they drive on—are not public but rather private goods.[4] Thus, their diffusion into society is more based on individual decision making than was the diffusion of timekeeping.

The first mechanical transportation devices were steam powered, appearing in England in the 1750s. In 1801, Richard Trevithick built the colorfully named Puffing Devil. One of the first in the United States was Oliver Evans's Orukter Amphibolos ("amphibious digger"). Although these devices were lumbering and awkward to use, the inventors certainly get high marks for their inventive names.[5] Evans's contrivance combined a small steam-driven boat with the wheels of a wagon. It is reported that in about 1804 he drove around Philadelphia and then charted a route down the Schuylkill River to its confluence with the Delaware River, then back up the Delaware to Philadelphia (Rakeman 1976, 25).[6]

By the 1880s, in Germany, Gottlieb Daimler had perfected the four-stroke gasoline engine, and Carl Benz had developed a commercially feasible automobile (Bryson 1994, 197; Flink 2001, 11). It was over the next two decades that the automobile industry really got on its feet.[7] It is fair to say, however, that these developments were met by a somewhat unreceptive world. If we think of a device being useful, then necessary, and finally taken for granted, cars were only beginning that cycle at this stage. In the late 1800s, none of the major elements of today's automobile culture were in place (Elliott and Urry 2010). From the 1880s until the 1910s, the automobile was a plaything for the rich or the technically resolute. It was not widely useful or an item of daily necessity (Flink 2001, 27). Cars cost their owners bundles of money to own and maintain. In addition, they needed constant maintenance and repair to keep them running. The earliest versions had no cabs for passengers. Further, there was simply no infrastructure to support their use. For example, there were no gas stations, so fuel had to either be bought by the bucket at a livery stable or ordered by the barrel and kept at home (Jackson 1985, 256).

As if to ensure cars' marginal status, the roads were poor (Urry 2007, 114).[8] Those that existed were often impassable (Howe 2007, 41; Rose 1976; Snider and Sheals 2003). In 1909, only slightly more than 8 percent of roads in the United States were surfaced. Moline notes, for example, that before the surfacing of roads, mud could be several feet deep during the springtime thaw in Illinois. Because of this, the maximum range of transportation previous to the automobile was less than twenty-five miles (Moline 1971, 25). It is with good reason that people who went away to school or married into a neighboring town only 5 to 10 miles distant rarely ventured back to their hometown (Moline 1971). Maps, guidebooks, or signs posts were few (Urry 2000, 61). The intrepid driver needed to rely on local advice or good luck when navigating cross-country.[9] Further, there were very few of the institutional artifacts we associate with automobile culture. Insurance policies, driver's licenses, and car registrations were nonexistent (Bryson 1994, 199; Urry 2007, 116). Meanwhile, the complex of settlement patterns, marketing, and policymaking was still set in its preautomotive phase, which meant no allocations for roads and other infrastructure.

This is not to say that no forces were pushing for the development of the car and car-based mobility. Cars were becoming more economical and efficient than horse-based transportation. In addition, the automobile brought about public health benefits, reduced rural isolation, and played into ideas of freedom and individualism (Elliott and Urry 2010) as well as notions of family unity (Flink 2001, 28). The issue of hygiene was a major drawback of horse-based transportation in the cities. According to James Flink (2001), before the automobile era, in New York City horses deposited 2.5 million pounds of manure and 60,000 gallons of urine every day. When dried and blown into the air, this entered the respiratory tracts of the citizens and carried with it a wide variety of ailments. In wet weather, it became a syrupy mess that was tracked into homes and workplaces. If this was not enough, Flink notes that each year in New York City 15,000 horses literally died in their tracks. This meant that on any given day about 40 to 50 newly dead horse carcasses started the process of rotting on the streets of Gotham (Flink 2001, 136). It was against this background that the automobile was introduced. Although it has its own forms of excrement and decay, these were not as obvious in the early 1900s as they are today.

Given the potential benefits of the car, the stage was set for the transition away from horse-based personal transportation. In 1898, for every 18,000 citizens of the United States, there was approximately 1 automobile. This changed dramatically during the next two decades (Jackson 1985; Kihlstedt 1983). In the fall of 1908, Henry Ford started to production of the Model T. Its price,[10] production methods, ruggedness, and relative ease of repair meant that at one point, 40 percent of all cars in sold in the United States were Model Ts. Eventually, automobiles began to gain general acceptance. By the 1920s there was a dramatic drop in the number of inhabitants per registered car (see figure 4.1).

The automobile was quickly adopted between 1910 and 1930 in many nations. The transition took place somewhat earlier in the United States, which did not see the same disruption of production facilities caused by the First World War in Europe. By the 1960s many industrialized nations had about 1 car per 4 to 6 people. According to Fischer, in the United States the price of an automobile fell from 14 percent of a manufacturing worker's earnings in 1914 to only about 2 percent a decade later (Fischer 1992, 111).

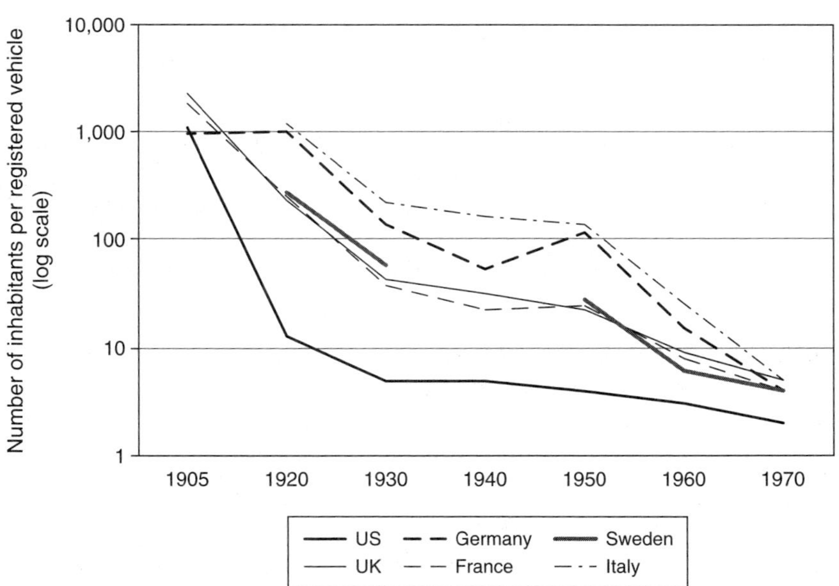

Figure 4.1
Number of inhabitants per registered vehicle for several nations, 1905–1970 (from Jackson 1985).

By mid-century, there were about 2 persons per car in the United States.[11] By comparison, by 2005 in Western Europe, there were just about 2 persons per car and in Eastern Europe about 4 persons per car.

The United States has a high number of automobiles per capita (75.5 per 100 persons in 2009) in comparison to other countries. France, for example, was ranked number 10 with 59.5 per 100 (Millard-Ball and Schipper 2010). In addition, people in the United States drive the cars that they own more than people in other countries. According to work done by Lee Schipper at the Lawrence Berkeley Labs in California, people in the United States drove over 20,000 kilometers per year and used public transportation for approximately 1,000 kilometers per year (Schipper et al. 2001).[12] In Denmark, people drove slightly more than half as much (about 12,000 km/person/year) and used public transport approximately three times as much (3,000 km/year/person). In the United States private cars account for about 95 percent of all transportation. By contrast, in Denmark public transportation approaches a quarter of the transportation used. These numbers also underscore the difference in settlement patterns and in the willingness to maintain a viable public transportation system.[13] A person in China travels about 1,000 kilometers per year, half of which is by public transport and another 40 percent of which is done by either walking or biking (Schipper 2009).

Analysis has shown that a perception of dependence on the automobile arose quickly in the diffusion process:

> by the Depression a family without a car faced special difficulties in satisfying its transportation needs: It took longer to reach relocated services in village centers by horse and wagon; the barns, liveries, harness shops and blacksmiths had dwindled as their owners converted them to auto dealerships, garages, gas stations and parking lots. As a result of these and similar changes, the car became a rural necessity. (Interrante 1983, 97)

Michael Berger (cited in Fischer 1992, 190) makes a similar point. With the coming of the auto, other social resources that had been oriented toward horse-based transportation were reduced or removed.[14] Because of this, the automobile became a necessity. During the Depression, people often prioritized the automobile over other expenses, as it was a way of securing work (Interrante 1983). This trend has continued. As of 2006, slightly more that 91 percent of Americans thought that the automobile was not something they could live without (Taylor, Funk and Clark 2006; see also Hjorthol et al. 2007).

By any measure, the automobile is a fixture on the scene. It is widely diffused. Not only in the United States, but in many countries around the world, we find a plethora of automobiles. Like the clock and the mobile phone, the car has moved from being perceived of as an oddity to being useful and finally to being critical for many people.

A "$1,200 Studebaker with a California Top": The Legitimation of the Automobile

Accompanying the diffusion of the automobile was a legitimation—as well as a counter-legitimation—process. This consisted in the development of narratives justifying—or opposing—the adoption of the car. We may see it as a necessary evil or use it with resignation. However, at one level or another, billions of people have drawn on these legitimations and justified the use of the car.

There have been many positive narratives associated with the car. The automobile has been described as a sign of modernity, a tool for pursuing jobs, and an agent in determining how we want to live and with whom we want to associate. It is a status symbol, it has been seen as a way to help maintain family solidarity, and quite often it is simply seen as a necessity.[15] All of these issues, and many more, are used to justify the car (Flichy 2007) and to help us make sense of our use. They place the use of the automobile into a broader sense of how society should be (Berger and Luckmann 1967). We are in general agreement that it is okay to have a car and use it to carry out our car-based activities. We might have the sense that we drive too much, are too ostentatious with our choice of car, or do not use the car appropriately. In spite of these problematic issues, we have developed a justification structure.

In its early days, the car was met with suggestions that horse-based transport was more reliable. However, by the period between the World Wars, the car came to be seen as the hallmark of modernity (Sanford 1983; Silk 1983; Urry 2000, 61). A car speeding along an open road became the metaphor for progress.

> The new highways were . . . not only a measure of the culture's technological prowess but were also fully integrated into the cultural economy. They were talked about as though they had an important democratizing role: the idea was that modern highways allowed more people to appreciate the wonders of nature. (Urry 2000, 72–73)

Cecil B. DeMille said that the car illustrated "the love of motion and speed, the restless urge toward improvement and expansion, and the kinetic energy of a young, vigorous nation" (cited in Hey 1983). According to Marinetti, "We declare that the splendor of the world has been enriched by a new beauty—the beauty of speed. A racing car . . . Racing along like a machine gun is more beautiful than the winged victory of Samothrace" (cited in Jackson 1985, 171). With the rise of the automobile, there were changes in clothing fashion and in the sense of what constitutes modern appearance (Frye 1983). The design of the automobile itself came to been seen as a type of art form (Guillory 1983); it now serves as a totem with which we live out different phases of our lives (Setterberg 1983) and through which can recount our past (Smith 1973).

According to another line of narration, the car helped maintain the unity of the family. Cars provided a space where the family could be assembled, on either a Sunday trip or a longer vacation, and share a common experience (though these commentators clearly did not envision siblings arguing in the back seat) (Fischer 1992, 248; Hey 1983; Lynd and Lynd 1929; Urry 2007, 113). Commentators also noted that the family could move out to the suburbs to avoid the buffeting of raw urban life. We see this in the Babbitt-like oratory of Russell Herman Conwell, who said:

> My friend, you take and drive me . . . out into the suburbs of Philadelphia, and introduce me to the people who own their homes around this great city, those beautiful homes with gardens and flowers, those magnificent homes so lovely in their art, and I will introduce you to the very best people in character as well as in enterprise in our city, and you know I will. (cited in Jackson 1985, 50)

Using a somewhat more critical point of departure, Mumford examined the reasons that motivated people to move to the suburbs, and consequently into a life that was more dependent on the automobile. The reason to move out of the city was:

> to be your own unique self; to build your unique house, midst in a unique landscape; to live in this Domain of Arnhem and self-centered life, in which private fantasy and caprice would have license to express themselves openly, in short to withdraw like a monk and live like a prince—this was the purpose of the original creators of the suburb. They proposed in effect to create an asylum, in which they could, as individuals, overcome the chronic defects of civilization while still commanding at will the privileges and benefits of urban society. (Mumford 1963, 6)

Both Conwell and Mumford saw that the car/suburban complex potentially gave the individual a sense of mastering his or her situation. It offered the idea, or in Mumford's terms the illusion, of individual fulfillment.

The automobile was a gauge of status. In their study of Middletown, Robert and Helen Lynd describe how it was (and still is) a way to assess others' social position (Lynd and Lynd 1929, 89–91). The people they interviewed were quick to note that they owned a particular type of car, such as the women who said that they owned a "$1,200 Studebaker with a California Top, semi enclosed" (ibid., 29). This car had allowed them to take an extended vacation and to see, among other things, Niagara Falls. These considerations were woven into a way of justifying ownership and use.

This is not to say that ownership was not contested. While the car was clearly a marker of local standing, the Lynds also noted the discussions of how using money on cars was seen as wasteful. Using the car too much could indicate poor judgment or mark the individual as a carouser (ibid., 149–150). The Lynds describe the conflicts between parents and their teenage children over the use of the family's car. They also discuss morally questionable driving on Sundays when people might have better been in church. During this period, we also gained the ideas of joy-riding and drunk driving (Flink 2001, 159). Going well beyond the pale of proper activity, the Lynds also note the use of the car as a setting for courting and sexual trysts, a theme that was picked up by many other authors (Flink 2001, 159–161; Lewis 1983).[16] Indeed, according to the Hoover Commission's findings, "the car helped loosen family ties, reduce parental authority over children." On the positive side, the Commission noted that it helped to "introduce women to new opportunities for recreation, romance and work outside the home; and in general expand social contacts between the sexes" (Sanford 1983). Thus, while one line of commentary suggested that the car brought the family together, another said that it had exactly the opposite effect.

In spite of these conflicting ideas, we have legitimated the automobile. We see it as the normal way of organizing our lives. Indeed, it is difficult to think of doing without. When asked, for example, whether they could have managed daily life without a car, a woman in a 2005 interview in Norway said flatly that "It wouldn't work." Another woman said, "I wouldn't make it to work [without the car]," and a third simply said, "I am dependent on it." It would not be possible for these women to work out their lives without cars. When imagining a day without a car, a Norwegian male named Georg fumbled his way to an answer:

> It is, it, it ruins all the time frames. So it, it depends . . . A day without a car, you just have to think about it and decide that today I am going to do as well as I can and get done what I can get done but it might affect others. So that if I have to take the kids there would be very little working and little of the normal tasks.

Georg and his family live in a suburban part of the greater Oslo area that is served by public transportation. Thus, he does in fact have other options. That said, he has arranged his activities and his way of conceptualizing daily life around access to the car.

We have developed a whole set of narratives that warn about the problems of the automobile. In recent times, for example, we have become concerned with it as a source of pollution; in earlier times, some people described it as potentially dangerous and as a locus for mischief. These ideas battled with the opposite line of commentary describing how the automobile was a harbinger of modernity and facilitated family unity. We also justify use by simply noting that it is not possible to deal with daily life without one. We note that the structure of modern society dictates use—that we are, in some sense, hostages to the structure of automobile-based society. There is also a social element in our legitimation structures. Georg comments that not having a car "might affect others." The comments of Georg and the others noted above also show that our justifications for using the car circle around the ideas of efficient action and fulfilling our social responsibilities. We look at this idea more closely in the next section.

The Impact of the Car on the Social Ecology

As noted above, in the period following the early 1900s, the automobile diffused into society. It moved from being a marginal contrivance to being a central element in our everyday lives. Along the way, it also fundamentally changed our social interactions and social institutions. According to Urry, "The car system re-organizes time and space and thus how people negotiate the opportunities for, and constraints upon, work, family life, childhood, leisure and pleasure" (2007, 117).

It entered into what Kenneth Boulding might have called an open niche. Before the large-scale production of oil and gas there was only steam and draft-animal-based land transportation (Boulding 1991, 4). The consequence of this is that society was organized in a different way. Stores, jobs, and schools were within walking distance. Kenneth Jackson (1985, 15) has written that, in even the largest US cities, only 1 person in 50 traveled more than a mile to his or her workplace. In a similar vein, Elliott and Urry note that in 1800 people traveled 50 meters a day; in 2010 they travel 50 kilometers (Elliott and Urry 2010, 2).

Those who commuted by train or a tram were constrained by its timetable mentality. This changed, however, with the arrival of the car and its flexibility (Urry 2000; 2007, 119). This transition also affected the distribution of services. Flink (2001) describes the situation in Ostrander, a small town in central Ohio, northwest of Columbus, near US Highway 36. Between 1890 and 1970, the population remained at about 1,500 to 1,600 people. In 1920, the main business area boasted fifteen establishments plus an opera house. The businesses included grocery stores, barbershops, a bakery, a grain elevator, a bank, and a drug store. By 1970, this had been more than halved since most of these functions, including the grocery store and the physician, were now located in the county seat fifteen minutes' drive away. Clearly, massive economic forces were at work. The logistics of maintaining the businesses in the small town were too complicated. Businesses were centralized and consolidated to realize efficiencies of scale. More recently, a Walmart Supercenter opened up thirty minutes away in the northern suburbs of Columbus. This again threatens the commercial basis of the remaining stores and facilities. These changes, however, would not have been possible without the automobile.

These transitions also reveal changes in the social ecology of transportation. The car out-competed other forms of transportation and occupied a central niche. This changed the dimensions of the niche to fit the dimensions of the car culture. Instead of driving rickety contrivances over muddy roads, we built more elegant and reliable devices, and we paved the roads. We have readjusted our living patters, the way we work, shop, entertain, and school ourselves. We have developed massive logistical systems to supply stores and shopping areas centered on truck-based distribution and automobile-based shopping. Thus, the population of automobiles did not reach some equilibrium level for the niche and stop there. Rather, we have progressively allowed the car to take over adjoining transportation niches. In all of this, the automobile has moved from being an oddity to being an essential element in the social ecology of transportation.

Balloon-frame Homes, Housing Subsidies, and Recalibration of Marketing: The Automobile and the Social Ecology of the City

According to Elliott and Urry (2010), the automobile is the most powerful of mobility systems. As we've seen, the transition in automobile ownership also corresponded to, and indeed helped cause, the growth of the suburbs (Crawford 1994; Dyckman 1973; see also Fischer 1992; Flink 2001, 129; Hall 1996; Hayden 2003; Jackson 1985; O'Connor 1982; Thorns 1972). By the early 1900s, the expansion of the city had begun with the development of urban rail systems (Jackson 1985, 101, 113). The trajectory of this development changed with the adoption of the automobile and road-based transportation. A parallel development was the use of trucks to make deliveries to manufacturing facilities and stores (Levinson 2011, 83–93). Previously, factories and stores often clustered near railheads in the center of the city to facilitate transportation. The flexibility of truck-based transportation and extensive systems of paved roads allowed factories, warehouses, and stores to move to more suburban locations. It also follows from this that jobs were moving out of the cities and into the surrounding areas (Interrante 1983, 92; O'Connor 1982).

Reduction in the costs of housing construction also played a role in the rise of the suburbs (Hayden 2003). Specifically, the so-called balloon-frame house revolutionized construction practices (Boorstin 1965, 148–152). Until the mid-1800s, houses were built using heavy timbering or simply using logs (or other materials) stacked atop one another. Access to standardized lumber and, as Jackson notes, generous use of machine-made nails facilitated this development. Building older types of homes required many different skills and a variety of hand tools with which to form, for example, the tendons and joints. The balloon-frame home, however, reduced the number tools needed and dramatically simplified the process (Jackson 1985, 127; Sprague 1981).

It has also been noted that the landline telephone played into the expansion of the city (Calhoun 1992). Jean Gottman has written, "There can be no doubt . . . that modern telephone systems, with their use of wires and waves, switchboards and computers, cables and satellites have made the space they serve more fungible for communication purposes. It became possible, in principle, for individuals located anywhere in that space to converse with one another" (Gottman 1977, 307). He continues, "The generalization of the individual motorcar and of the telephone have actively aided suburban sprawl" (Gottman 1977, 312). The landline telephone facilitated the needs of intimates and trusted friends to communicate and spawned the development of telephone-based service scheduling (e.g., calling a doctor's office, lawyer's office, barbershop, or automobile repair shop for an appointment) and it also allowed for the management of economic affairs.[17]

Starting in the 1930s in the United States and accelerating after the Second World War, federal housing subsidies also favored the development of the suburbs (Jackson 1985, 206–215). In many cases, it was less expensive for a middle-class family to own a home in the suburbs than to rent a smaller apartment in the city center. This also had the effect of turning the core cities into increasingly unattractive low-rent districts without stable residents. In addition, accounting practices that allowed for the accelerated depreciation of new buildings also encouraged the growth outside the core city areas of office buildings, strip malls, and shopping centers, where the suburbanites could find jobs (Hayden 2003). The end of the Second World War and the subsequent baby boom added further pressure for the expansion of the suburbs and the Interstate Highway System (Rakeman 1976).[18]

Before the Second World War, a host of smaller builders were responsible for a patchwork of developments. After the war, however, building was dominated more by a small number of large-scale builders who carried out much more extensive projects. The Levitts, who developed Levittown on Long Island, are often seen as the prototypical example of mass development (Gans 1967). They employed scales of production and standardized building practices to assemble homes inexpensively and quickly. According to Jackson, they were building more than thirty homes a day at the peak of their production. He also notes that on a single day in 1949, the organization was able to draw up 1,400 contracts with families who had been waiting their turn to buy a home (Jackson 1985, 235). In all of this, the US federal government was actively supporting the development through various financial programs that gave the middle class the wherewithal to buy homes. From the 1880s to 1940, about 40 percent of homes were privately owned. After World War II, federal regulation and housing subsidies pushed this up to about 60 percent, where it remains today.[19]

Construction methods, communication systems, and transportation systems merged in the form of the suburbs. With this development, the automobile came into its own. When compared to trams and railroads, the automobile gave people greater control over their movements. It allowed the individual to choose his or her own schedule (Elliott and Urry 2010, 328; Jackson 1985). Thus, the stage was set for the expansion of the cities beyond their traditional boundaries. The automobile became a common part of everyday life.

The impact of this settlement pattern was a growing reliance (perhaps even overreliance) on the automobile for personal transportation. The car was moving from being simply useful to being a necessity. Urban developments grew beyond walking distance (Cowan 1983, 83; Flink 2001, 3; Foster 1983; Lynd and Lynd 1929, 64; Urry 2000, 51), resulting in what Dolores Hayden calls the "drive to lunch syndrome" (2003, 156). Not surprisingly, in a comparative study of Muncie, Indiana—the "Middletown" studied by the Lynds in the 1920s—Bahr and colleagues found that people were more often working and enjoying recreation at a greater radius outside the home than suggested by the earlier work of the Lynds (Bahr et al. 2004, 266; see also Schroeder 2007, 112).[20]

Los Angeles is often cited as a city that has fostered the use of the automobile and in turn has been formed by its use (Flink 2001, 142). The annexation of the San Fernando Valley in 1915 and its subsequent development—along with the plotting and intrigues of those seeking access to the needed water rights (Reisner 1986)—allowed for the extensive expansion of the suburbs and their single-family homes. During the decades after this annexation, thousands of developments were started and hundreds of thousands of single-family homes were built. According to Jackson, by 1930, 94 percent of all dwellings in Los Angeles were single-family suburban homes (Jackson 1985, 179). Consequently, Los Angeles never developed the same type of city core seen in cities that developed before the prominence of the automobile (Flink 2001, 142–144).

The increasing reliance on the automobile changed the social ecology by closing off other transportation options and changing how Los Angelinos (as well as suburbanites in other similar cities) socialize. In many cases, trams and trolleys were out-competed by the car. In other cases, they were bought up by the interests of automobile companies and sold overseas—as in the case of the streetcar system in Los Angeles being bought by General Motors and sold to Hong Kong (Jackson 1985, 171). By the start of World War II, many millions of Americans lived beyond areas served by public transportation. Access to jobs, school, recreational activities, social engagements, shopping, and just about any other activity we can think of relied (and still relies) on the car (Crawford 1994; Thorns 1972). In addition, we see the development of massive systems that support the need for a car.[21] We often base our vacations on driving to automobile-friendly hotels, eating fast food, and enjoying automobile-accessible sights and attractions. Even if we never use an automobile, the wares that we purchase are usually delivered using the automobile/truck-based transportation system. In short, we are living in a drive-in society.

The "Drive-In" Culture

By the 1960s, the development of the automobile/suburban complex took on a sense of inevitability (Urry 2007, 114). Indeed there was a sense of path-dependency. "The [automobile based] lock-in means that specific institutions structure how this system developed." According to Urry (2007, 116–117), the importance of the automobile can be seen in the development of gas stations, strip malls and a broad variety of "drive-in" businesses. Starting out as a begrudging extension of livery stables, the gas station soon took on dimensions of grandeur. In 1905, the enterprising C. H. Laessig of St. Louis connected a garden hose to a hot-water-heater tank and created the first gas pump. By the 1930s, a motorist could buy gas from a building designed to appear as a Greek temple, a lighthouse, a Chinese pagoda, or an art-deco confection. From there gas stations morphed into more standardized locations that eventually included small stores (Jackson 1985,

256–257). Today, they are not just gas stations but service areas where we provide for the nutritional needs of both the car and the passengers within. In some cases, these service stations have become a central source for culture in the locality, since we can also buy music, kitschy art, a quick meal, and reading material.

According to Ruth Schwartz Cowan, the adoption of the car and the development of truck-based transport and the refrigerator led to different types of food consumption (Cowan 1983; see also Levinson 2011, 125–126). Prior to the automobile, the logistical system for supplying stores was based not on road transport from distant warehouses but on local production supplemented by some goods transported by rail. In addition, home-based food preservation (canning, making preserves, and salting meat) and food preparation (baking, etc.) were more prevalent. With the rise of the automobile culture, fewer things were made in the home and more were made in factories and transported by truck to local stores. The Lynds, for example, describe the commercialization of food production, clothing, cleaning, and other products (Lynd and Lynd 1929). One example is bread. The Lynds note that in 1890 about 25 percent of bread in the United States was commercially baked. By 1930 in the United States, bread was store-bought in 66 percent of rural and 75 percent of urban homes (Interrante 1983, 98). The result of this was (and is) more of an emphasis on fresh wares and less use of home-based canning and preserving of food. This leads to the consequent need to continually replenish stocks in the home and in the store, both of which in turn rely on car/truck-based transportation systems (Cowan 1983, 174).

Inspired by the possibilities of marketing aimed at the motorized public, a variety of "drive-in" offerings were soon available (Flink 2001, 161). Hotels moved from being located in the center of the city near the railway station to being along the highway in the form of cabins, and finally to becoming motor hotels or motels (Belasco 1983; Flink 2001, 183).[22] Drive-in restaurants serving fast food first appeared in the 1920s and have since begun to allow patrons to not only drive in, but to drive through. Not to be outdone, some states allow for drive-through liquor stores. When the demons are done with their raging, there are drive-in places of worship, and when we face that ultimate cul-de-sac, there are even drive-through funeral homes.[23]

Another feature of the contemporary landscape that has been shaped by the automobile is the shopping center with off-street parking. In the 1920s, builders started to cluster stores together and surround them with easily accessible parking areas. The posh Country Club Plaza in Kansas City was the first explicitly automobile-accessible shopping center. The development included fountains and Moorish-style towers. The key development, however, was that it prioritized automobile over pedestrian accessibility. By the 1950s, the shopping area was becoming an enclosed, climate-controlled world in itself, and by 1980, shopping centers accounted for the large majority of retailing in the United States (Jackson 1985, 259). This in turn led to the development of the supermall. The size of some shopping malls is measured in millions of square feet. Today they include hundreds of stores and a wide variety of entertainment offerings.

The path development of the automobile/suburban system has reorganized. Urry writes that it "coerces people into an intense flexibility," which he compares to Weber's iron cage (Schroeder and Ling, forthcoming; Urry 2007, 120). Urry writes:

> The car is immensely flexible and wholly coercive. Automobility is a source of freedom, the "freedom of the road," because of its flexibility which enables the car driver to travel at speed, at any time in any direction, along the complex road systems of western societies which link together almost all houses, workplaces and leisure sites. Cars therefore extend where people can go to and hence what they as humans are able to do. Much social life could not be undertaken without the flexibilities of the car. But at the same time such a flexibility is coerced, it is necessitated by automobility because the moving car forces people to orchestrate in complex and heterogeneous ways their mobilities and socialities across very significant distances. Automobility necessarily divides workplaces from the home producing lengthy commutes; it splits home and shopping and destroys local retailing outlets; it separates home and various kinds of leisure sites; it splits up families which live in distant places; it necessitates leisure visits to sites lying on the road network; it entraps people in congestion, jams and temporal uncertainties and it encapsulates people in a privileged, cocooned, moving environment. (Urry 2000, 60)

For many of us, our homes, work, leisure activities, shopping, and schooling are all geographically separated, connected by a system of roads and parking that compels us to use the car. There are in some cases other options (e.g., subways in New York or bicycles in Copenhagen and Amsterdam), and indeed, sometimes the alternatives are better suited to our needs. However, without question, our use of the automobile has reorganized the geographical (and social) dispersion of many cities (Elliott and Urry 2010, 52). Approximately a third of the space in London and something like half of Los Angeles are given over to exclusively car-based activities. The car/suburban complex is perhaps one of the most massive constructions of humanity. It has many dark sides, but we are seemingly fated to coexist with it in spite of the resources it requires and the damage it does.

These car space areas exert an awesome spatial and temporal dominance over surrounding environments, transforming what can be seen, heard, smelt and even tasted (the spatial and temporal range of which varies for each of the senses). (Urry 2000, 193)

This transition in the social landscape described by Urry speaks to the way the car has changed the social ecology.

"My Best Friends Live in the City but by No Means in This Neighborhood": Reciprocal Expectations

We have largely accepted the logic of the automobile. Our thoughts concerning transportation assume access to and use of the car. In addition, the car has reformed and often constrained opportunities for sociation. With our more-or-less willing acceptance, the automobile has restructured society in its own image. It has become a system that facilitates sociability (Urry 2007, 116).

The question remains as to how the automobile mediates social interaction. We saw that clock time has become a tacit way in which we organizing our social lives. A single, widely accepted system governs how we keep track of time. This lets us assume, when making agreements, that all the people involved will have a common way to coordinate their interactions. Everybody will know when to be there, and everybody will know who is early and who is late.

For many but not all, the car has become synonymous with transportation. The car changed many of the expectations associated with work, friendship, access to services, and leisure patterns. These activities are no longer based on propinquity (Flink 2001, 153; see also Novak and Sykora 2007; Gordon and de Souza e Silva 2011, 110).[24] Flink reports on a survey of residents in a section of Los Angeles in 1929. The survey found that

a new definition of a friendly neighborhood is apparent; it is one in which the neighbors tend to their own business. One householder explained: "We have nothing whatsoever to do with my neighbors. I don't even know their names or know them to speak to. My best friends live in the city but by no means in this neighborhood. We belong to no clubs and we do not attend any local church. We go auto riding, visiting and uptown to theaters." The interviewer concluded, "This straining against the bonds that hold them in the area makes for many families an uneasy, unsettled, uncertain state." But another observer of the 1920s Los Angeles interpreted the phenomenon positively: "The role particularly of the automobile . . . has cut down spatial distance and tended to increase social nearness to such an extent that every person may live in wide-flung communalities of his own [making], in place of the old closely circumscribed neighborhoods." (Flink 2001, 168)

Urry also notes that the automobile has changed the way that "People . . . assemble complex, fragile and contingent patterns of social life, patterns that constitute self-created narratives of the reflexive self" (Urry 2007, 122).[25] The size of modern cities has had consequences for the maintenance of relationships and the ability to provide support to others (Mok and Wellman 2007).

Whether or not the car is tearing at the local social fabric, it allows or perhaps demands that we work out our social interactions across greater distances. By the mid-1920s, the car had become more reliable and its price had fallen. It was possible for many families to own a car. As the social geography was reformatted to fit the range and possibilities of the car, we began to operate on a larger map.

The car also changed our mutual expectations. Many jobs, social engagements, and leisure time activities operate on the presumption that we can transport ourselves by car. Employers, for example, do not pay for transportation. The assumption is that the individual pays for this out of his or her own pocket. When arranging social activity (e.g., a soccer game for a child or a meeting of the local bluegrass band), the assumption is that all the participants arrange for their own transportation in the vast reaches of suburbia. It may be a car pool or some sort of turn-taking, but there is the assumption that the participants will work this out. A friend who lives in an adjacent suburb might invite us to a party. The invitation, however, does not assume that the friend will supply the transportation. It is assumed without comment that the invited partygoers are able supply their own means of getting there. The same set of assumptions is true of jobs, shopping, education, hobby groups, and our children's after-school activities. This was not as salient an issue in the preautomobile world since the geographic range of social interaction was not as large.

If the car is not available, we have to prioritize and take into account who will be inconvenienced or disappointed. Lienert (1983) tells of the situation of her mother:

> My mother [who did not drive] . . . had to depend on someone else to take her whenever she wanted to go anywhere. And since my father often didn't wish to go places, my mother spent a lot more time at home in her mature years. . . . The automobile, rather than freeing her, confined her more than ever.

Where there are alternative forms of transportation available, this is not a problem. Buses, trams, subways, and bike paths (not to mention compact housing arrangements) reduce the need for private automobiles. However, the interaction between suburban development and the diffusion of the automobile mean that such substitutes are not always available. In these settings, it is difficult to imagine carrying on our daily life in the absence of the automobile. Thus there is the assumption that we can provide for our own transportation needs.

In suburban geographies, the individual without a car has to throw him- or herself onto the charity of those who do. People without cars have a limited radius of action and limited ability to find work (Ehrenreich 2008; Hjorthol 2000); indeed, people who do not have a car and are in search of work have a significantly reduced the chance of getting a job (Allard, Tolman, and Rosen 2003). This issue also has a gendered aspect. In his analysis of Levittown, Herbert Gans notes that women were particularly interested in getting cars in order to minimize isolation (Gans 1967; see also Chen and McKnight 2007; 1967). Car ownership facilitated shopping and supported the logistics of various family activities. According to Ruth Swartz Cowan, the car was the "location" where she could be most often found (Cowan 1983, 85), although women often have a more limited travel radius than males (Hjorthol 2000; Lenhart et al. 2007).

The "Constraining Power" of the Car Culture

As we have seen, timekeeping and the automobile have achieved a certain stability in society. Timekeeping is a part of the social framework that we have lived with for several centuries. Indeed, the social development of mechanical timekeeping has been reified to the degree that it is often difficult to see it as a social construction. It has assumed a crystallized form that is not malleable in any real sense (Berger and Luckmann 1967, 76). By contrast, the automobile has been with us for a little over a century, and in that time, it has reformed how we socialize and the way we organize our lives. The system of the automobile/suburbs has achieved a facticity.

> The car is thus not simply an extension of each individual; automobility is not simply an act of consumption, because of the way that it reconfigures the modes of sociality. Social life has always entailed various mobilities but the car has transformed these in a distinct combination of both flexibility and coercion. (Urry 2000, 190)

To be sure, the car is not benign. It is indeed a threat in many ways. Car accidents cause death and maiming. The system occupies vast tracts of land that were once available for food production. It induces a level of stress into daily life as we rush hither and yon, and, perhaps most importantly, it is polluting the planet (Elliott and Urry 2010, 188). In spite of these issues, the car is a central technology of social mediation.

Both timekeeping and car-based transportation systems have established themselves and achieved a certain taken-for-granted role. In each case the development of this role has had a formative stage. Both clock time and the car/suburban system have passed a critical point where they have moved from being a curiosity to where they play a central role in the way that we organize our social interactions. They have engendered a legitimation system and rearranged the social ecology.

The car and the clock are not necessarily immutable. While both are thoroughly embedded in the social fabric, it is possible to conceive of changes in how we use timekeeping and transportation. Indeed, as we will see in chapter 9, the mobile phone has had some influence in this area (Ling 2002). The mobile phone, and the direct interpersonal access that it provides, can be seen as a challenge (or perhaps a supplement) to the coordination dimensions of timekeeping (Ling 2004). Nonetheless, it is far-fetched to say that the clock will disappear from society. To use Boulding's metaphor, there may be adjustments in the niche, but it will continue to occupy the niche.

In a similar way, the automobile has occupied a position in society and has relentlessly out-competed other forms of transportation. The car gives us freedom to move, but it has also given us various social problems (Urry 2000, 189). The massive investment in roads, highways, but also in buildings (homes, shopping centers, truck-based warehouses, etc.) in many—but not all[26]—cities makes us dependent on the car. In the vast suburbs of cities like Los Angeles, London, Denver, and Sydney, there are few alternatives to the car. In many suburban cultures, not to have a car is to not fully participate in society (Urry 2000, 191); hence the car has become a technology of social mediation. Having passed a critical point whereby cars were seen as a necessity, the automobile/suburban complex has taken on a tangible facticity. Like timekeeping, people living in the suburbs cannot ignore the exigencies of the car. In the words of Durkheim:

> Even when in fact I can struggle free from these rules and successfully break them, it is never without being forced to fight against them. Even if in the end they are overcome, they make their constraining power sufficiently felt in the resistance that they afford. There is no innovator, even a fortunate one, whose ventures do not encounter opposition of this kind. (Durkheim 1938, 51)

Urry echoes Durkheim when he says that automobile provides us with freedom while also constraining us (Urry 2007, 117). We are jostled into accepting the role of the automobile in daily life. It means that our friends and our social life are more distant. However, the structure of the urban landscape is such that we have to accept its presence. It has become one of the ways in which we mediate our social interaction.[27]

5 "If I Didn't Have a Mobile Phone Then I Would Be Stuck": The Diffusion of Mobile Communication

Clock time gives us coordination, sometimes at the expense of stress. The car gives us mobility in the vast frontiers of the suburbs. The mobile phone gives us access to one another. For many, mobile telephony has gained the perception of having reached a critical mass. It has moved from being a plaything for rich business people to being an assumed part of our collective lives. It is nearly ubiquitous among teens in some countries (Vaage 2007; Leonard and Sensiper 1998). In short, the mobile phone has rearranged the social furniture of our experience.

Many of us have grown to assume that others are perpetually available. Others take for granted that we have a mobile phone. It is important, however, to note that the diffusion of the mobile phone is at different stages in different countries. In Europe, North America, and Oceania, for example, the device is, for all intents and purposes, ubiquitous, and advanced smartphones are increasingly common. In many developing countries, the mobile phone is only starting to be widespread. In this chapter, I trace the diffusion of mobile communication and its attainment of a critical mass. Chapter 6 takes up the question of how we justify using the mobile phone; chapter 7 examines how the mobile phone has actually changed the social world nesting in society; chapter 8 discusses how we increasingly have mutual expectations of access; and finally, chapter 9 examines the mobile phone in comparison to the car and timekeeping.

The Technical Foundation of Mobile Communication

Mobile communication has grown from being a minor element in the broader telecommunication landscape to being a central element. Indeed, texting and calling are among the most common forms of mediated interaction in many cultures (Farley 2005; Lenhart et al. 2010). In addition, the use of social networking and location-based services is growing (Sutko and de Souza e Silva 2011).

The mobile phone is one of the minor miracles in our technical lives. It was mentioned as almost an afterthought in the so-called Maitland report on telecommunication in the developing world from the mid-1980s (Maitland 1984). Since that time, it has spread around the globe. The cellular system of mobile telephony grew from the foundation provided by the landline telephone and a variety of other radio-based systems (Agar 2003, 35; Farley 2005). Starting in the 1920s, mobile radio (though not cellular communication) was used in various public sector services (police, fire) and some more commercial services (taxis). During World War II, radio was used for the coordination of military operations. At this point, radio-based communication and the traditional landline-based telephony systems were separate. Aside from special instances, it was not possible to make a call from a telephone and patch it over to the radio-based system or vice versa. After the war, there was some movement in this direction. Experiments were done to examine whether radio could be substituted for wired systems for people living in rural areas. The reasoning was that it was far cheaper to use radio signals than to string a wire out to farmers living miles from the nearest town (Ling 2004).[1]

The main research that resulted in the cellular network was conducted at Bell Labs in the late 1940s. More efficient data processing and the notion of the cellular system were essential elements of this research. The transistor—developed by John Bardeen and Walter Bratton with the cooperation of William Shockley in 1947—provided more efficient data processing. Indeed, the transistor is at the base of the broader technological revolution that has resulted in the PC, the internet, the smartphone, and just about every other digital device that we can think of. Like the vacuum tube before it, it radically reduced the energy needed to carry out various operations. In the case of mobile communication, it allowed for, among other things, the computing power to follow many thousands of mobile handsets as they move across the landscape.

The other major element was the cellular architecture for the telephone system developed by Douglas H. Ring and W. Rae Young in 1947. Up to this point, the approach to radio communication had been to use a single antenna in a central location intended to cover a large area such as a whole city. This approach was technically easier, since it required only one set of transmission equipment. However, it did not make efficient use of the limited radio frequency. The beginnings of civilian mobile telephony came in the postwar period. A single-tower "broadcast" system allowing drivers to call from cars was developed in the New York/Boston area. It was based not on the cellular model but rather on the broadcast approach that only allowed a limited number of channels. The late 1960s saw an increasing demand from industry for more advanced mobile communication that could service more customers.

D. H. Ring and W. R. Young suggested in 1947 that each coverage area should be geographically divided into a mesh of cells that would allow for the reuse and the sharing of relatively limited radio capacity.[2] This grid of smaller cells was far more efficient since one frequency was not monopolized by a single cell for a whole city; rather the frequency was reused many times in different cells that were remote from one another. AT&T developed a trial cellular system that was used on the Amtrak Metroliner train between Washington, D.C., and New York in 1969. As the train moved from one "cell" to another, calls were handed off from one tower to the next. Four years later, Motorola developed a handheld mobile phone, and in 1977, a cellular system was established in Chicago that included ten cell towers and allowed for 2,000 users.[3]

To digress slightly, mobile communication employs radio frequencies that are higher than often used in earlier applications. This gives the system the ability to carry reasonable amounts of data, but it limits the frequency range. In radio terms, not all frequencies are born equal. The frequency refers to the number of radio waves, or hertz,[4] that pass a given point in a given amount of time. This can range from portions of one cycle to many billions per second, eventually moving through the range of visible light and beyond. A basic characteristic of the lower-frequency radio signals is that they can bear their message through almost any impediment. Indeed, some very low radio frequencies can penetrate almost anything. This means that they can carry their message over long distances without any dampening of the effect. These signals will gladly rush off around the globe to deliver their message to the awaiting receiver. The problem with these frequencies is that there are not many oscillations or waves and so they cannot carry much information. Thus, although they can be spread far and wide, they cannot say much.[5]

The opposite effect is true of the high-range frequencies. The higher the frequency, the less willing it is to penetrate obstacles. Light, which is an extremely high frequency, cannot, for example, pass through walls. Thus, the high radio frequencies cannot be counted on to carry their signal over hill and dale to far-off locations. Rather, they are dampened as they spread. Indeed, with some forms of high-frequency communication, the signal cannot penetrate trees or the occasional rain shower. For much higher frequencies, the sender and the receiver need something approaching a "line of sight" connection. In their favor, however, the higher frequency radio waves are plentiful and can carry a lot of information. The oscillations of the radio waves (each wave is called a hertz) moves up through the tens and hundreds to thousands (kilohertz), millions (megahertz), billions (gigahertz), and trillions (terahertz) of radio waves per second. Beyond that are the peta, exa, zetta, and the positively Jedi-sounding yottahertz ranges.[6] The early Nordic Mobile Telephone (NMT) system was introduced at the 450-megahertz frequency (i.e., 450 million radio waves per second). In radio communication terms, that is a modest number.

As the NMT system for mobile communication became more popular, a second frequency was used, the NMT 900 system (900 million waves per second). Seen from this perspective, the movement up a couple of hundred megahertz is an exceedingly small portion of the total radio spectrum, but the phones in the 450 and the 900 systems had quite different characteristics. The message from the 450-megahertz towers carried further than those of the 900 system. In a country like Norway, with large tracts of sparsely inhabited land, rural people noted the difference. In particular, hunters and residents of remote cottages knew that they could—sometimes with only minor gymnastic ability—count on a connection with the 450 system, whereas the 900 system was more dodgy (Ling, Julsrud, and Krogh 1998). As the more advanced Global System for Mobile Communication (GSM) became available (in the 900- and the 1800-megahertz range), there was political, commercial, and technical pressure to decommission the 450 system. Other applications required the spectrum, it was expensive to maintain the equipment, and there was a movement toward the GSM system at the higher frequency. In addition, more users could be accommodated with the new system. However, the decommissioning of the 450 system was met with reluctance and bitter comments by the rural population. Thus, the seemingly simple technical issue of reallocation can have social effects.

The technical parts of the mobile phone system include the mobile handset (also called the terminal)[7] and the system of towers and exchanges. The mobile handset is perhaps the most complex bit of machinery that we commonly carry with us. It includes a radio receiver and a sender. In addition, it has a display and an ever-increasing amount of computing power with which to surf the net and administer apps of all types. All of this is packed into a device so small that we can lose it in the sofa cushions.[8]

Beyond the actual handset, there is a vast network supporting mobile communications. As noted by Colin Cherry (1977) with reference to the landline phone, to focus on the handset and ignore the "back office" part of the system is to disregard a major part of the technology. This system is more complex than that of the landline telephone. Where the landline system was satisfied to efficiently direct the calls through a switching system (either a manual system or, later, on an automated one) to a particular number, the mobile telecommunication system has to keep track of the location of all the phones on a system in the event that a call is directed to any of them. This is done by sending a radio signal to the handset every few seconds just to find which cell tower to use in case there is a call. If, by chance, the person making or receiving the call is moving, the cell network also needs to work out how to hand over the call to the adjoining cell so that there will be no interruption in the call.[9] With these many technical pieces in place, the mobile phone system was poised for adoption by society.

The Diffusion of Mobile Communication

In previous chapters, we have seen how timekeeping and the automobile were diffused into society. In each case, the technology developed in a particular social context and was nurtured for some time before it got a foothold and then gained broader diffusion and reached critical mass.

In the case of mobile telephony, as we have seen, things were beginning to happen in the late 1960s and early 1970s. Systems were starting to be developed in the United States. A few people were also starting to use the first mobile telephones in Europe at this time. Svein Johannesen (1981), who collected the market research reports carried out by Norwegian Telecom between 1966 and 1981, describes market research on nascent mobile telephony along with the stirrings of other new data-related services such as paging, "electronic mail," fax machines, market liberalization, and colored landline telephone handsets with button-based dialing. Interestingly, research was also still being carried out on marine communication (a significant enterprise in Norway) and telegraph messages. Indeed, we read that in 1978, based on a sample of 1,042 persons, 29 percent had sent a telegram in the past year.[10] Thus, while traditional forms of communication were still on the agenda, many different communication possibilities were beginning to emerge.

In the late 1970s, the mobile phone seemed to meet the needs of a smallish niche of people. It took some time, however, before it was perceived as a commonly available gadget. The earliest version of the mobile telephone in Norway (not based on the cellular system) was commercialized in December of 1966 (Bastiansen 2006). In 1967, the number of subscribers rose from 41 in January to 149 by December. During that period, the number of calls went from 527 in January to 2,646 in December. In 1975, ten years after the introduction of the mobile telephone, there were approximately 10,200 subscribers (about 2 percent of the population) who generated 1.4 million calls (Bastiansen 2006), largely by businessmen (and they were specifically men).[11] In the United States during the period of the commercialization of the cellular network from 1983 to 1991, there were 5.1 million users (Rakow and Navarrow 1993) (as in Norway, about 2 percent of the population).

The telephone handsets from this period were heavy and solidly built. A typical device was the 1966 Nera CM 13V, which weighed 9.35 kilos (20.5 pounds) and cost as much as $1,000. Photos of these devices show that they had a combination microphone/speaker and a variety of dials that are not a part of today's mobile phones, including "squelch," "channel," a "call" button, and a function dial that allowed for, among other things, "receive," and "stand-by" (Nordsveen 2006). They were clearly intended for people who were not technologically timid and who could bear a heavy burden, both physically and financially. According to Johannesen (1981), a 1975 survey answered by 1,350 mobile phone users in Norway shows that for 84 percent of the respondents, the main reason they bought a mobile phone was that it fulfilled needs for communication during working hours. In addition, 40 percent of the respondents noted that they used it in work-related communication outside working hours. Interestingly, 3 percent noted that one of the reasons they bought it was for security. As we will see below, this became a primary justification for purchasing a mobile phone as it became more popular. In 1975, approximately half of the Norwegian users were salespeople, and another 25 percent were associated with transportation services. In a 1980 survey also reviewed by Johannesen (1981), of 234 mobile phone users interviewed, 65 percent had their mobile phones installed in automobiles, and 12 percent were located in a private home or cottage. The analysis shows that 23 percent were "mobile" in the sense that they were not bound to any one location or to a car. At this point, mobile communication was a niche product. Its diffusion was clearly far short of critical mass.

The pager was the immediate precursor to the mobile phone. It was quite popular before the widespread adoption of the mobile phone, and perhaps it alerted people to the possibilities (and the stress) of being contacted regardless of where they might be. Pagers were more difficult to use, since upon receiving a page the person had to find a landline telephone and call the person who had sent the message.[12] A study from 1976 among 1,516 Norwegians (Johannesen 1981) ranked interest in pagers behind services such as specified bills, push-button telephones, and the ability to assign buttons to often-called numbers on landline phones. The ability to be reached by others when not near a landline phone was clearly not on the minds of Norwegians at this point. The following year, 596 persons were asked about pagers. Specifically, they were asked if they would prefer a pager or a mobile phone. The analysis shows that 40 percent of the respondents preferred a pager while only 19 percent preferred a mobile phone. From our perspective, this seems like an odd choice. However, the mobile phones of the time were, as noted, large, bulky, expensive, and difficult to operate. These considerations perhaps account for the popularity of the smaller, simpler, portable, and far less expensive pager. Johannesen (1981) reports that in 1980, there was an attempt to predict the number of mobile phones needed in Europe. The analysis suggested that there could be a market for 20 million subscriptions in the near future (i.e., about 2 percent of the European population at that time). The analysis also suggested that about half of the phones would be mounted in cars and the remaining terminals could be portable.

The dynamics changed in the early 1990s in Europe with the commercialization of the GSM (Bakalis, Abeln, and Mante-Meijer 1997). By 2008, there were 199 million subscriptions in Russia, 150 million in Germany, 90 million in Italy, 77 million in the UK, and 50 million in France. Indeed, there were 20 million subscriptions in the Netherlands alone. It is clear that back in 1980 there was no clear sense as to how ingrained the mobile telephone would become.

The regulatory history of mobile communication has both encouraged and hindered the diffusion of mobile phones. In the United States, the diffusion was rather slow (Abrahamson 2003). The Federal Communications Commission (FCC), which is responsible for regulating mobile telephony, broke up the massive Bell system as mobile technology matured. In the spirit of local competition, the United States was divided up into over 400 different license areas. The intention was that two competing mobile companies would be granted a license in each of the areas. Clearly, the most attractive license areas were the large cities. The FCC intended to invite operators to seek a license and to grant them based on a review of the qualifications of the groups applying. When the bidding process was opened for the top markets, however, the FCC was flooded with applications (Agar 2003, 41). There were 92,000 applications for the top 180 metropolitan areas. The FCC then switched to a simple lottery system, and, as might be expected, many of the applicants had no technical understanding of mobile communication. This resulted in a disjointed patchwork of operators, technical systems, roaming functionality, and subscription types. In effect, this poorly regulated Klondike approach, along with odd pricing policies, handicapped mobile communication in the United States for many years.

By contrast, Europe still used the system of large public operators. In many cases, it was simply decided to deploy a single mobile telephone system for an entire country. The Nordic countries (Norway, Sweden, Denmark, and Finland) developed the NMT system that allowed for international roaming. Somewhat later the European Commission developed the above-mentioned GSM system that allowed for international roaming, texting, and a separation between the handset and the subscription (represented by the SIM card). By the mid-1990s, GSM covered most of Western Europe. It has continued to be the dominant mobile technology, serving about 85 percent of all subscribers as of this writing. In Japan, a sort of compromise emerged between the private development as seen in the United States and the public standards of Europe. This resulted in several consortia, all of which shared a common technical standard. This also meant that there was the same basis for interoperability in the Japanese system as in Europe (Agar 2003).

Worldwide Adoption

Mobile telephony has followed an exceedingly fast growth curve. In many places, it seemed to have reached critical mass by the turn of the millennium. As we have seen, there were only a relative handful of mobile phones in the mid-1980s. In 2000, there were just fewer than one billion landline phones and about 750 million mobile phones; by the start of the next decade, there were about 4.6 billion mobile phone subscriptions, while the number of landline subscriptions was falling. Seen in other terms, in 2000 there were 12 mobile phone subscribers and 6 internet users per 100 persons on the plant. By 2010, there were 78 mobile phone subscriptions and 30 internet users (and 13 mobile broadband customers) per 100 persons. The popularity of mobile telephony can be seen in the arc of the upper curve in figure 5.1 (ITU 2009).[13]

In the late 1990s, the mobile phone began to establish itself as a popular form of mediation in many countries. At that time, operators began to realize that if they subsidized the cost of buying a handset, it would generate traffic. They were willing to sponsor the cost of handsets so long as the user was "locked" to the operator for a given period. Thus, new subscribers were able to buy mobile phones for nominal prices. The cost of subsidizing the phones was recouped during the period of the contract.

As of 2009, the country with the largest number of mobile phone subscriptions was China with 747 million subscriptions. China was followed by India (525 million), the United States (298 million), Russia (230 million), and Brazil (174 million). Indonesia, Japan, Germany, and Pakistan round

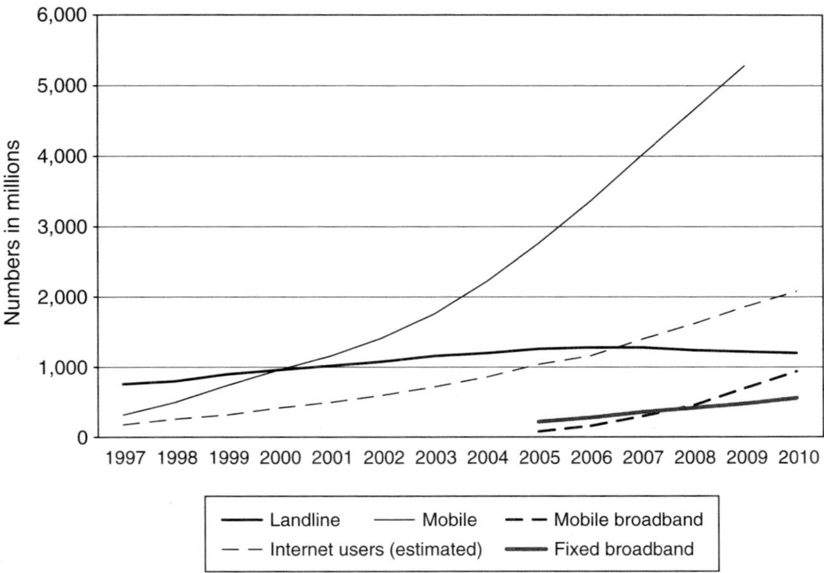

Figure 5.1
Number of landline phones, mobile phones, PCs, and internet users from 1997–2010 (source: ITU).

out the group of countries with more than 100 million subscriptions. Naturally, countries with small populations would have few total subscriptions. In the Marshall Islands and Kiribati there were 1,000 subscriptions while Tuvalu had 2,000 in 2009.

The number of subscriptions per 100 persons allows us to see the density of mobile telephony in a country and reveals a different picture. Tuvalu had 28 mobile subscriptions per 100 inhabitants. Thus, in spite of the fact that it has very few phones totally, it ranks only somewhat behind India (43 per 100 inhabitants). As of 2009, the number of subscriptions per 100 persons ranged from near zero for some countries to more than 200 for some others. The countries with the greatest density of subscriptions include the United Arab Emirates, Bahrain, Macao, and Antigua.

Moreover, to these relatively affluent countries we can add several countries that do not often figure as in such rankings. Estonia, Panama, and Dominica are among the countries with the highest density of mobile phone subscriptions. Dominica, for example, ranks as number 12 in terms of mobile subscription density even though it has a relatively moderate Gross Domestic Product (GDP) per capita. In spite of these seeming barriers,

Dominica has a robust mobile communications sector with 159 subscriptions per 100 people, and, as in other countries, the number of landline subscriptions is dropping. There are a small number of internet subscriptions in Dominica, about 10 per 100 people.[14]

How is it that this tiny country has found a place among the countries with the highest per capita mobile phone subscriptions? There are different elements to the answer. The mobile phone is clearly an accessible technology. Unlike the PC and the internet, the mobile phone does not require a large personal investment to buy the handset and connect to the network. Used or pirate handsets can be bought (Qiu and Wallis 2009). Prepaid subscriptions often are quite inexpensive and easy to establish. Indeed, people in developing countries often have several subscriptions (Kalba 2007) since different providers have different coverage and pricing profiles. One provider might have lower prices during the daytime, while another might be cheaper in the evening. One provider might have better coverage in Roseau, the capital of Dominica, while another has better coverage on the north coast of the island. Since prepaid subscriptions are easily acquired and inexpensive, people often have several of them and switch the SIM cards in and out of their phone as the need arises.[15] Another factor that drives up the number of subscriptions are those that are sold to tourists. It is, for example, possible to buy a small prepaid subscription for less than $5 that allows for incoming calls, texting, and local access. A tourist with an appropriate phone can buy a subscription for use while in the country and then simply throw it away when returning home. Finally, operators might be slow to remove unused subscriptions from their books.[16]

The case of Dominica shows that adoption can happen in countries with only moderate means. This is not to say, however, that money does not matter. Looking somewhat broadly at the international adoption of mobile communication, we find different adoption phases (see figure 5.2). After the commercialization of GSM, a group of relatively affluent countries adopted quickly. The top 20 percent of countries rated by GDP per capita include the United States, many Western European countries, the affluent Asian countries, and members of the OECD. In 1999, nine countries had more than 50 subscriptions per 100 people (ITU 2009). The Nordic countries were among this group. They were flush with the success of the early NMT system, and the GSM system was quickly adopted. In addition, Italy, Austria, Hong Kong, Taiwan, and South Korea were eager mobile phone users. By 2005, 18 countries had an adoption rate of greater than 100 subscriptions per person. Countries having the middle range of GDP per capita—many Central and South American and Eastern European countries such as Brazil, Ecuador, Serbia, Morocco, Mexico, and Moldova—did not start adopting mobile phones until after the "internet bubble" of 2000. However, by mid-decade they were quickly moving into the mobile phone era. The countries with the lowest GDP per capita were moving into the take-off phase at the end of the period covered in figure 5.2. These include countries such as Chad, the Central African Republic, Turkmenistan, Vanuatu, Yemen, Sierra Leone, Zimbabwe, and Somalia. The sheer weight of poverty hinders the development of a viable mobile communication system in such countries. In other countries, such as Cuba and Burma, political issues have hampered adoption.

The numbers described here show that mobile telephony has gained a foothold in many places. Indeed, the mobile phone is often seen as a necessity. If we speculate that a perceived critical mass of users is on its way to being established when one-third of the population own subscriptions, then approximately 75 percent of the countries in the world have reached critical mass. If we are somewhat more stringent and say that there must be one subscription for every other person, still about two-thirds of the countries examined by the International Telecommunications Union (ITU) met this standard in 2010 (ITU 2010). It is clear that, as shown in figure 5.2, those who have lower GDP per capita are less likely to reach this level. The main point remains, however: the mobile phone is widely diffused. It is also

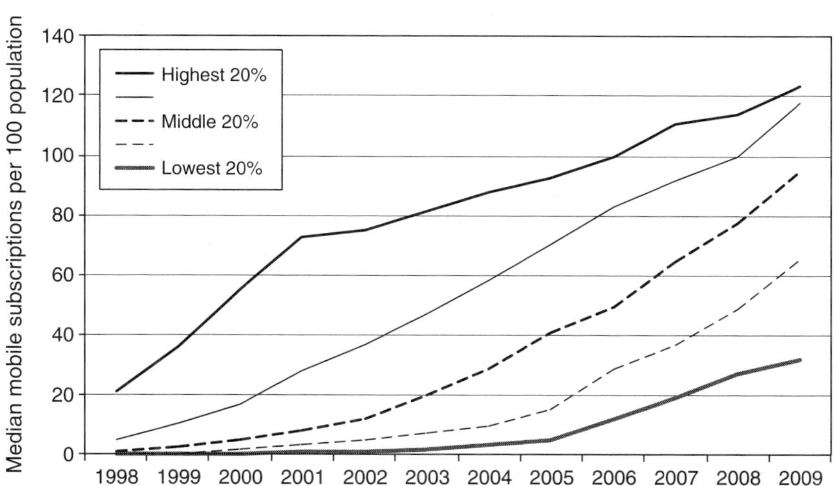

Figure 5.2
Adoption of mobile telephone subscriptions by GDP per capita, 1998–2009 (source: ITU).

possible to suggest that the perception that it is common for others to have a mobile phone is widespread.

Adoption in Norway

The broad-brush analysis of mobile phone adoption throughout the world gives us one perspective on the way that mobile telephony has diffused and become customary. It is also useful to look more carefully at how different groups within a country have adopted the device. Fortunately, there is a backlog of data documenting this for Norway. Indeed, we have data starting in the late 1990s that help us to see the way that mobile telephony spread through different groups in society. It is true that Norway is not a microcosm of the world; its experience is different from that of other countries. Norway is an affluent country that is politically stable and has a small population. It was one of the earliest countries to adopt the mobile phone on a broad scale. However, its experience also illustrates the progression from "mobile phone as a useful but contested object," through the development of critical mass to the stage of "mobile phone as necessity." Available data for this country include both qualitative and quantitative and go back to the mid-1990s.

As noted earlier, the first commercial service started in 1966. However, mass adoption had to wait until the early 1990s (Bastiansen 2006). For a time, Norway was one of the most advanced countries when judged by the number of subscriptions per 100 people. Not all age groups, however, were as quick to adopt the mobile phone. The contours of the adoption process can be seen in data for Norway from 1997 to 2005 (see figure 5.3). The data for teens show the progressive diffusion of the mobile phone.[17]

In 1997, about four years after the commercialization of GSM, mobile communication started to take off among teens.[18] In particular, older teens were starting to adopt the device. This was before the advent of prepaid subscriptions, so having a mobile phone represented a significant economic commitment. This helps to clarify why younger teens did not have mobile phones at that time. They were less likely to have jobs and the economic wherewithal to have a phone. It is also worth noting that this was before teens' discovery of texting. Two years later, in 1999, the picture was quite different. While older teens still led the way, there had been a transition in ownership among younger teens. By 2001 the large majority of older teens reported having a mobile phone, and about half of younger teens reported the same.

The analysis shows that ownership had become very common for the majority of Norwegian teens. Indeed, after 2005 we find nearly ubiquitous

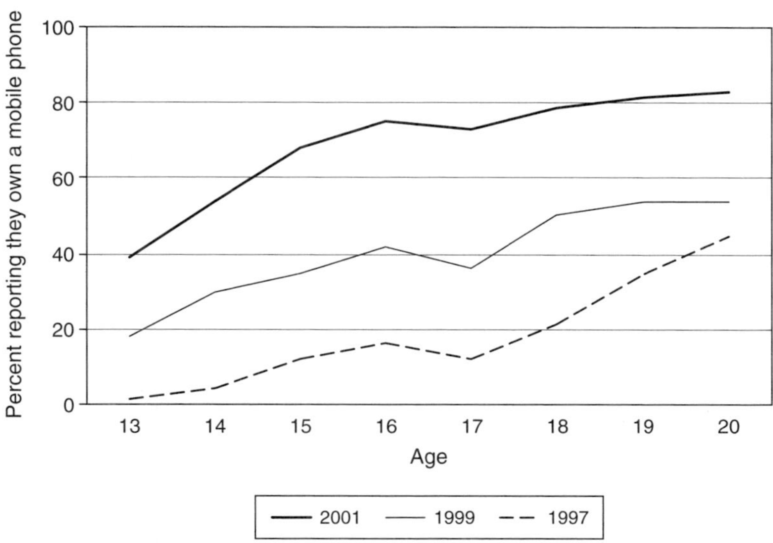

Figure 5.3
Percent of Norwegians reporting who owned a mobile phone by age, 1997–2007.

ownership among teens and young adults. In 2005 and 2007, these data from Norway's national statistics bureau, with its sophisticated sampling procedures, failed to find any people between the ages of 17 and 30 without a phone. For people between the ages of 30 and 70, the reported ownership rate was well over 90 percent (Vaage 2008). According to the ITU material, there were about 116 subscriptions per 100 persons of all ages in 2007 in Norway (ITU 2010).

The introduction of prepaid subscriptions and the discovery of texting[19] helped to push the adoption of the device among teens. As more teens acquired a mobile phone, it increasingly became a status symbol and the locus of social interaction. Pressure increased on those who did not have one to join the network. In some few cases, teens adopted an antitechnology stance (Ling 2004). However, this cut them off from certain interactions with their friends and family. The functionality of mobile telephony, along with its position as a cultural icon, meant that the device was soon perceived as being omnipresent, at least among teens.

Similar trends are seen in other parts of the world. The mobile phone is nearly ubiquitous in Denmark (Denkel and Madsen 2010); a 2009 survey of parents of teenagers in the United States noted that 73 percent of all teens aged 12 to 17 have a mobile phone (Lenhart et al. 2010). The survey also reported that 90 percent of the parents have one. Indeed, in the United

States, about 1 home in 7 has only a mobile phone connection and no landline telephone (Blumberg and Luke 2007).

In Norway, the mobile phone gained the sense of having a critical mass by the mid-1990s. A similar though perhaps time-shifted pattern has been found in many other countries. It has become so thoroughly embedded in our daily lives that in many countries—and more importantly in many groups—it is taken for granted. The numbers outlined above give the statistical picture of diffusion. Nonetheless, it is important to understand that it is our perception of diffusion that is at play (Lin and Hsu 2009; Lou, Luo, and Strong 2000). There is the increasing sense that most people are available via their mobile phones. The holdouts will perhaps be pressured to get a mobile phone by their friends and family. To be sure, those people who hold out must increasingly have a justification near at hand for not having a mobile phone (Ling 2008) (e.g., it induces stress, it is expensive, or there is danger from electromagnetic waves).

For the users, however, the widespread acceptance of the device means that it is easier to be in contact. Some modified version of Metcalfe's law (discussed in chap. 2) suggests that we collectively gain value from the system since we are able to call, text, or use mobile Facebook with others. In other words, the more people there are who own a mobile phone, the better we are able to maintain social contact.

From Communication Monoculture to a Complex Personal Communication Environment

Coming first to more affluent countries, but spreading to almost all societies, the mobile phone has changed where and how we communicate. At the same time, a change has occurred in the way that we carry out interpersonal communication. The landline phone was, until the mid-1990s, the dominant channel in a type of communication monoculture. There was largely one channel for mediated communication, namely landline telephony. Now there are many, including email, blogging, micro-blogging (i.e., Twitter), instant messaging, texting, and social networking, in addition to voice interaction via the mobile phone, the landline, and services such as Skype and Viber. Some of these are quasi-broadcast while others are more often focused on one-to-one interpersonal communication.

Figure 5.4 shows the relative daily use of different forms of commonly used mediation technologies among Norwegian teens in 1997.[20] The landline telephone was clearly the main form of electronic mediation at this time. About 8 in 10 teens reported using it on a daily basis. Interestingly, there was a slight tendency that use was declining among the oldest teens. At the same time, mobile voice interaction was reported by about a quarter of the 19- and 20-year-olds. The use of pagers was a part of the scene at that point. Teens had developed elaborate pager codes with which they could agree on meetings and inform others of their plans. Pagers and pager subscriptions were inexpensive and within the reach of teens (or at least within the reach of their parents). Figure 5.4 shows that in 1997 the pager was losing popularity when compared to mobile voice. To be sure, within a few years, pager systems would be discontinued.

Several forms of mediation are not represented in figure 5.4. Texting, for example, was technically possible, but was not being used by teens—or anyone else, for that matter (Trosby 2004; Hillebrand et al, 2010). Indeed, texting was only popularly "discovered" in the period immediately after the survey was carried out.[21] It is also interesting to note that email was the only net-based form of interaction at that point. It had only a marginal position in the communication ecosystem.

Figure 5.5 shows that in 2008, the mediation landscape in Norway was quite different from a decade earlier.[22] Landline telephony, mobile voice, and email were still on the scene, but the landline was no longer the dominant form of interpersonal mediation. In 1997, approximately 80 percent

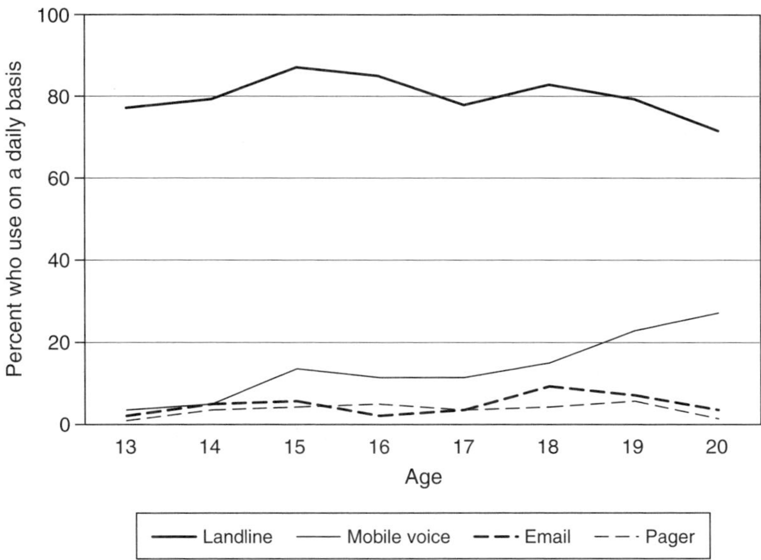

Figure 5.4
Daily use of different mediation forms by Norwegian teens in 1997.

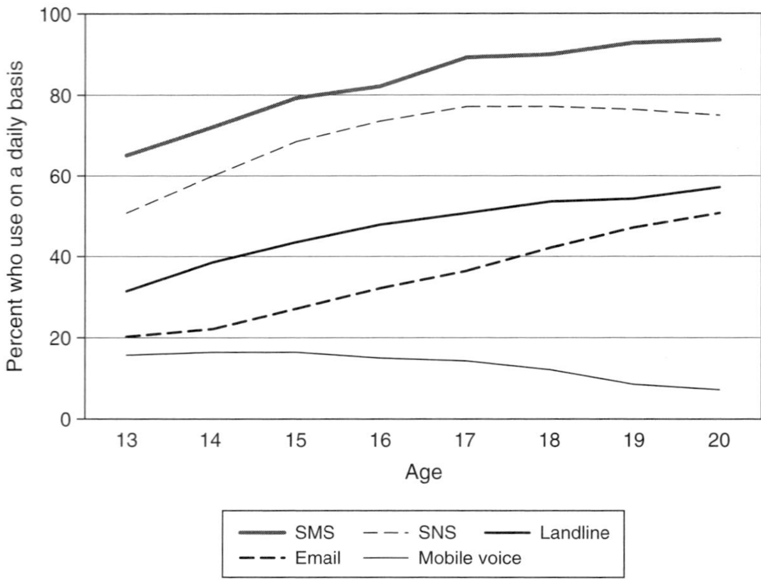

Figure 5.5
Daily use of different mediation forms by Norwegian teens in 2008.

of teens used landline telephony on a daily basis. By 2008, this was dramatically reduced. Norwegian teens had experienced the transition from the landline to the mobile phone and were keenly aware of the freedom that the mobile phone offered. This comes out in the comments of teens in a focus group from this time:[23]

Erica: I live out in the country and so it is very important for me to be available. Because most of the people I know drive a car and they drive past where I live and if I didn't have a mobile phone then I would be stuck. They usually call me and ask if they should get me. Another thing is that when you are going to meet people then you have to get in touch with them and most people don't sit at home every day, so it pays to have a mobile phone.
Interviewer: Yeah, you don't sit at home; can you say how you make arrangements? Why is a mobile phone so important?
Erica: Because it is how you get in touch with people you are going to make arrangements with. They never sit at home and I don't either.

These teens are operating in a social sphere where there is a critical mass of mobile phone users. For them it is useless to wait for a call to come to a landline phone. The mobile phone gives them a sense of independence and access to one another. They describe the central position of the mobile phone in their lives. These comments also help us to understand why teens were so quick to drop the use of the landline phone. Being bound to a physical place—most notably their homes that were decidedly not in the center of the action—was something to be avoided. The mobile phone was the key to their social world.

Looking again at figure 5.5, email had gone from being used by less than 10 percent of the teens to being somewhat more widely used by older teens. Interestingly, mobile voice use had stayed about the same at 13 percent for all teens. In 1997, older teens reported more use of mobile voice. In 2008, the youngest teens reported the most use. In 2008, pagers had disappeared; indeed, pagers are no longer supported in the telephone network. The most significant changes in 2008 were texting and social networking sites (e.g., Facebook, MySpace, and Flicker). Over 90 percent of the oldest teens reported texting on a daily basis in 2008. These developments came onto the scene and gained widespread popularity.[24]

The contrast between 1997 and 2008 shows us the rise of texting as a form of personal mediation. It was quite literally unused in 1997; a decade later, it had become a central element in the media landscape. Indeed, in many countries and among some groups, texting is the most used form of mediation (Ling 2010; Ling, Bertel, and Sundsøy 2012). Data from the United States also show that it has been widely adopted there. In the United States, text messaging has become the most common way for teens to reach their friends. According to data from the Pew Internet and American Life Project, texting surpasses face-to-face interaction, email, instant messaging, and interaction on social network sites as the most common way that teens are in contact with their friends. In 2009, a typical teen in the United States sent and received 50 texts a day for a total of 1,500 per month. As many as 31 percent of teens sent and received 100 per day (Lenhart et al. 2010).

The other new mediation form shown in figure 5.5 is that of social networking sites (SNS). As with texting, these were not a part of the media landscape in 1998. At that point, there were multiuser domains (MUDs) and MOOs (MUDs, object-oriented), but the web 2.0 forms of social networking would not appear for several years. In 2008, just under 70 percent of Norwegian teens aged 13 to 20 reported using an SNS on a daily basis, with a peak in use among 17-year-olds (77 percent).[25] Two years later in 2010, more than 80 percent of teens in Norway reported using a social networking site (Vaage 2011). Since 2010 the use of SNS has continued to grow.

These trends are not unique for Norway. Others have noted the same transition. In a report on U.S. teens, Ito et al. note that "Young people in the United States today are growing up in a media ecology where digital and networked media are playing an increasingly central role" (Ito et al.

2010, 30). They are using a mix of mobile and PC-based communication (Livingstone et al. 2011), and they are using it in one-to-one interaction as well as in quasi-broadcast mode.

Looking at this through the lens of mobile communication, we see a move toward more net-based use of mobile phones (Ling and Sundsøy 2009). Analysis shows that a growing portion of all users, not just teens, own internet-capable phones. In Norway as of 2012, 27 percent of the population owned a "high-end" smartphone, that is, a phone with an advanced operating system, a powerful processor, and a touch screen. Another 25 percent had a more modest smartphone, one lacking a touch screen and powerful processor. It is tempting to think that these phones are taking over all mobile communication, and in many countries that is true (Zickuhr and Smith 2011). However, in Bangladesh, smartphones of any type are almost nonexistent in any age group: only about 8 percent of the phones there are classified as smartphones. One-third of the phones there are simple "entry level" devices, and another third of all phones are "unknown" pirate phones, usually produced in China.[26] This suggests that while traditional voice/text-based mobile phones have reached a critical mass, more advanced mobile phones are in the process of diffusing into society.

Conclusion

Mobile phones (and in some cases, advanced smartphones) have diffused into society such that in developed countries it is rare to find an individual without one. In the Global South it is often nearly as common. Coming on the heels of the great technical push associated with World War II, the small device we carry with us arose from the development of the transistor, the integrated circuit, and the cellular system of radio transmission. This groundwork, along with the development of different technical standards and diverse national regulatory regimes, formed and directed the trajectory of the mobile phone system as we know it.

Prepaid subscriptions and the subsidizing of inexpensive handsets encouraged the popularization of mobile telephony. These developments enabled people to use the system for increasingly mundane activities, which, in turn, contributed to the idea that a critical mass of people was available via the mobile phone. It was not only well-heeled business people that could afford one. Rather, the mobile phone was adopted by all sectors of society in many countries. This diffusion came first in the affluent countries, but it has followed progressively in those locations with a lower GDP.

Texting has also made an important contribution to the popularity of the device. Finn Trosby (2004), who was involved in the standardization of this service, called it the ugly duckling of GSM, since early on, many people did not see its worth (Trosby 2004; Hillebrand et al 2010). However, like the ugly duckling, it has been transformed. It is clear from our use of texting that we have a need for asynchronous, personal text-based messaging. We use it to make arrangements with others and we use it to gossip and joke with one another. Some of the functions of texting are now moving over to social networking sites.

In many cultures and in many groups, the perception has been around for some time that having a mobile phone is not just useful, but expected. In many social groups, the mobile phone is perceived to be nearly ubiquitous (Lin and Hsu 2009; Lou et al. 2000). As the number of users increased, the device spread rapidly. Holdouts then either were cajoled into buying a phone or they constructed an ideology of non-use (Katz 2008).

During the last decade, the mobile phone has morphed into a complex device. It is a calendar and a repository of personal information. It is a music player as well as the place where we keep our "to do" list and an app containing the family shopping list. It is a gaming device, an internet portal, and even a movie screen. We can take photos, compose love letters, create spreadsheets, check the weather, update our status, and read the news on our mobile phones. Increasingly, it is an access point to the mobile internet (Brown, Campbell, and Ling 2011), where we use it for social networking on sites such as Facebook.

In spite of all the changes and additions to the mobile phone, the use of the system still largely revolves around interpersonal communication. Analysis of traffic data shows that more than 90 percent of the events recorded consist of either texting or talking (Ling and McEwen 2010).[27] As much as anything else, the mobile phone represents a link between people, making communications more personalized, direct, and timely. This process has not happened without comment. It has been the object of both scorn and praise. Some have derided the loss of other forms of communication, while others have seen mobile telephony as the start of an exciting new way to communicate. Both of these perspectives describe the process of people locating the technology in their lives. They describe how, in the adoption process, we need to construct affirmations of the device in question. These help facilitate the diffusion and acceptance process by framing the justifications for having a phone. The legitimation—and derision—of mobile telephony is what I will now turn to.

6 "We Are Either Abused or Spoiled by It—It Is Difficult to Say": Constructing Legitimacy for the Mobile Phone

The mobile phone has become increasingly common. People often cannot think of leaving home without their purse or wallet, their keys, and their mobile phone. Many teens feel excluded from social life if they do not have a mobile phone. There is clearly a sense among some people that having a mobile phone is essential (Lenhart et al. 2010, 68).

While many people welcome the device, there are also those who are more reserved. Some see it as a talisman from the gods of technology, while others prefer a world without it. As the device has become interwoven into society, we have had to work out our understanding of just what the technology is and what is its raison d'être (Flichy 2007, 11). On a more practical level, we need to understand when and where we might use it, as well as the problems and issues associated with using it. These are the basic elements of the domestication approach (Sliverstone, Hirsch, and Morley 1992; Haddon 2003). Beyond that, we use justifications to support our use (or rejection) of technologies. These justifications often come from commercial sources touting the features of the device or system. The producers of mobile handsets, the network operators, and the content providers all have an interest in supporting the ideology of acceptance. However, our justifications of a technology can also arise in the process of daily interaction with peers (Berger and Luckmann 1967, 135).

A part of the justification system can also be seen in the development of an iconography. Like the Chrysler Airflow automobile of the 1930s, the mobile phone has become an icon of the early twenty-first century. We are interested in the form, design, and functionality of the new iPhone or Android. Themes such as these excite discussion around the lunch table at work. In addition to the fascination with the mobile phone itself, we have built up a series of good reasons to have one. These justifications can be seen as a type of broad ideological structure. We need to exemplify this broader structure with fresh illustrations so that the broader ideology does not become stale. Our narratives about security, accessibility, efficiency, and coordination are tightly woven into our personal justifications for having a mobile phone. To keep these justifications alive, we need to cultivate concrete examples of just how our mobile phones fit into and embellish these broader themes. We relate how we called a friend who saved us when we forgot our keys or how we were able to work out a complex meeting arrangement when the subway was late. These narratives fit into themes of, for example, efficiency and stress reduction.

A counter ideology also exists. People suggest, perhaps rightfully, that using a particular expensive mobile phone is "showing off," that mobile phones cause stress (not relieve it), that they are a threat to health, a disturbance, a breach of courtesy, and a waste of money. As with the pro-mobile lines of discussion, the people who hold this position need to supplement these points with fresh examples, or else the justification becomes stranded.

Thus, the mobile telephone is not merely a simple functional object. It is also the locus of a broader discussion. We love it and we hate it. It makes us happy when we get that long-awaited call or when we are able to save the day by performing some mobile coordination when we are in a pinch. But it is also the bearer of bad tidings, and it is expensive. There are two parallel lines of thought. A retired participant in a 1995 focus group,[1] Johannes, saw these different and contradictory dimensions:

> I wanted to have it with me when I was out walking, either in Nordmarka [the forest near Oslo] or in the mountains, to let others know if something has happened, so I could move about freely. It can be a little difficult to walk in Nordmarka on weekdays. Because at Kikut [a location in Nordmarka] and in the past there, there is not another person around, so that is the only reason that I got one. My son had one at the cottage and we have a lot of grandchildren and they had friends there, it [the mobile phone] was in the living room and it was used a lot, [calling] down to where they had friends, it wasn't free a minute I don't think. It must be a problem for people who have to put the mobile phone away. It costs money.

On the one hand, Johannes gives us the description of the mobile phone as a safety device that allows him freedom of movement, especially since he enjoys walking in the forests. At the same time, overuse, or use by people for trivial ends, is frivolous. Johannes sees that the mobile phone can facilitate safety but that it can be misused.

Johannes, in some ways, embodies the two narratives associated with the mobile phone. It is both a helpful communication tool and a wasteful diversion. In the battle between these two accounts, it is clear that the former has overtaken the latter. Although we like to carp about how stressful, expensive, or annoying the mobile phone can be, our general acceptance of the phone indicates that these are generally just words.

The mobile phone plays with our sense of what the sociologist Anthony Giddens (1991) calls ontological security—our sense of stability, continuity, and order. Following on Goffman's sense of ritual interaction as a basic element in society, Giddens says that we are able to order our lives by observing a set of orientations that are more or less patterned (Giddens 1991, 47, 82). So long as we feel that these patterns are in place, things are okay. Should they be disturbed, however, we need to work out how to deal with the new situation.

The widespread adoption of technologies of social mediation such as the mobile phone almost certainly disturbs the prearranged social situation. People use them in unexpected places such as on the street corner, and teens might use them to deceive their parents and to gain more freedom. Issues such as these spark the need for new legitimation work. We need to think through the potential upsides, and downsides, of the technology. As the technology emerges, is found useful—or not—and eventually becomes seen as necessary, we are involved in the process of justifying our use. The goal of this is a sense of constancy and order, what Roger Silverstone et al. called "a sense of confidence or trust in the world as it appears to be" (Silverstone et al. 1992, 16).

This is the case with mobile telephony. There are stories of how mobile communication is disruptive. At the same time, there are stories of how it is a useful tool and, in some ways, a necessity. We have stories, for example, of how that guy on the bus was talking too loud and saying ugly things on his phone. We also have stories of how we used the mobile phone to direct rescuers to a person in need and of how the device has assisted us in rearranging meetings or telling others that we will be late because of traffic. These stories describe how we are legitimating the different dimensions of ownership and use. They also connect mobile communication to established norms and values such as security and the need to manage our social interactions in a way that is accountable.

"I Have to Have One because I Am a Director on Several Boards": Early Mobile Ownership as a Status Indicator

There has been an evolution in our attitudes toward the mobile phone. Before its popular adoption, the mobile phone was mostly used by relatively wealthy people who had powerful positions (Agar 2003). The few people who used mobile phones often did so discretely, usually using a car-mounted device that might also, with some effort, be transported to vacation homes or other locations (Johannesen 1981). Ownership was also seen as a status indicator, as we can read in the comments of Jakob, an adult male from the 1995 Norwegian focus groups:

> Jakob: There was a big store that had mobile phones. They found the right one for us and we took it and were registered and then the bill came to 32,000 kr. [approx. $5,500]. We didn't believe it. We had forgotten to ask. We had that for many years and used it everywhere. Later we exchanged that for one that cost 15,000 [approx. $2,500] and then the small nice one that cost 3,500 [approx. $600]. So mobile phones are just an expense. But I have to have one, because I am a director on several boards . . .

For Jakob, ownership of a mobile telephone was a business expense, albeit a pricey one. It was also a status symbol. Few people are able to so casually discuss these sums of money. The assertion that he had to have one because of his central position in society marks him as a privileged person. Jakob's mobile phone was for the elite, a phenomenon also described by John Agar (2003) in the UK.

In the period of the 32,000 kroner mobile phone, there was no wide-scale public recognition of the need for mobile telephony. Calls were not made on the street. Restaurant patrons did not have them sitting on the table. Their use was not a public spectacle. The world existed in a premobile equilibrium. Phone calls were made from a landline telephone in relatively predictable locations (the hallway in the home, the office, or in a phone booth). At the time of the 32,000 kroner phone, telephony was generally a geographically bounded activity. We had already been through the exercise of developing our sense of etiquette associated with the landline telephone (Marvin 1988, 89).

As mobile telephony slowly edged its way into society in the early 1990s, it still carried with it the sense that it was associated with business and was a province of the rich (Nordal 2000). Indeed, during this period, many of the mobile phones were not nearly as portable as the "pocket format" phones that started to appear in the late 1990s. On those few occasions when we did see someone using a mobile phone, we were able to understand it as a slightly expanded version of landline telephony for a very limited number of business people. It was used to assist the functioning of industry. By mid-decade, this was changing. The increasing popularity of the device and its falling price meant that our sense of telephony was on the point of transformation.

"First Come the Show-Offs and after a While It Is OK": Negative Mobile Telephony Narrations

In the late 1990s, as an increasing number of people started to own and use mobile phones, many viewed the device with suspicion. Its use was often seen as showing off, vulgar, nerve-racking, and rude. When it diffused further, the mobile phone was seen as a threat to our sense of order. Rather than being used for important decisions in business or administration, it was being used for mundane interaction and pointless chitchat.[2] It was not only a status-intense tool of efficient management; it was also a sign of social decline. It was no longer the untarnished sign of rank, but became tinged with vulgarity. In addition, the mobile phone conflicted with the previous order associated with the landline telephone (Berger and Luckmann 1967, 124). As the mobile phone slimmed down in size and emerged from the car, use in public spaces contradicted the established sense of decorum. The response was to develop accounts of the mobile phone that outlined it as a threat to the social order.

One of the first reactions was to focus on the way that the mobile phone challenges our sense of manners. For many, its use was a breach of propriety. Users were portrayed as upstarts or pompous prigs who wanted to make themselves bigger than they were. Regardless of the factual basis, these stories helped us organize our thoughts about the mobile phone and helped to instruct us as to how we should behave when using the mobile phone. The stories about self-important mobile phone users who ignored etiquette and disregarded the feelings of others nearby were a brake on the acceptance of mobile telephony, albeit a break that did not hold.

"It Is Completely Screwed Up": Status Disruption and Inexpensive Mobile Telephony

As the industry matured, new groups started to use mobile phones. Rather than the businessman in a tailored suit, it might be the young hipster whom we were inclined to see as arrogant, since he was violating the common sense of modesty. A series of urban myths became associated with the mobile phone, pointing to its insecure position in our affections (Nesvold 2006; Grazioli and Carrell 2002). These stories reveal a sense that using the mobile phone in public places was a way to show that the user was affluent and well connected. However, according to the stories, these people were pretenders who were exposed. One version describes how a person—usually a man—was loudly talking on a flashy new device in a very public place. The story continues that an accident occurred nearby and the mobile phone user was asked to call emergency services—whereupon it was discovered that the phone was a fake.[3] Even though there is a subtext that using the mobile phone to call in emergencies is legitimate, this urban myth plays on the fact that talking on the mobile phone in public spaces was seen as a facade used to enhance status.

Material from the 1995 Norwegian interviews illustrates that people were sensitive to the idea that mobile phones were mostly a status item. One respondent, Arne, acknowledged this when he said:

> I really hope that it will not be cheaper to use [the mobile phone] because that would be a catastrophe. It is bad enough now. If it is so inexpensive that everyone can afford to use it, then it will not be so important to get out the phone when I am out to show that I have one.

Arne understood not only the functionality of the device, but also its symbolic status. For others, this sentiment was not as palatable. Some people in 1995 felt uncomfortable with the way that others played on the faux exclusive nature of the mobile phone.

> Peter: I think it is a fashion thing. So many of my colleagues, when they get out in the car, the first thing they take is the mobile phone, regardless of if it is blinking or whatever it is doing. It is a fashion thing. . . .

In Peter's estimation, his colleagues' use of the mobile phone was not too far from the situation depicted in the urban myth. The fact that they felt the need to "flash" the device in order to impress others was crude. It was not so much a functional activity as it was a way of asserting position and showing off.

Peter's comments on the popularization of mobile communication show the gap between accessibility and general acceptance. Up to that point, the cost of ownership and use had meant that the mobile phone was accessible only to the well-to-do (or to the pretenders). This, in turn, meant that it was easy to parody mobile phone users as upper-class show-offs. The situation, however, was changing. Falling prices and greater accessibility meant that mobile communication was no longer only for the rich, although the idea that mobile telephony was for everyone was still in the future. In a similar vein, the comments of Steinar, another participant in the focus group, indicate that teens were not seen as rightful mobile phone users:

> They walk around in the streets, young children, you know. They stand there with a mobile telephone. Someone said that they had taken a bus after Christmas and there was a group of boys that had been skiing and they sat there talking to their parents who were on their way to get them. They said that they would be here or there, you know. It is completely screwed up.

Steinar describes what has become one of the central functions of the mobile phone, namely, microcoordination. However, he does not see the teens' use as legitimate.

As adoption of the mobile phone spread into society, nontraditional groups started to have access to the technology. Some became suspicious if people who were not clearly associated with the business world used a mobile phone (Nordal 2000). The material from the 1995 focus groups shows that the respondents were not sure as to the motives of new user groups.

Sigmund: For my part, I maybe have a tendency to suspect people of showing off, if I see a 20-year-old with a mobile phone [more than if I see] a 40-year-old.
Knut: I think that there are more pushers who have mobile phones, criminals, real criminals.
Interviewer: That sounds serious.
Knut: You know, I come from Oppsal, and there is major criminality, and there are all the worst people there and they have mobile phones. They call all the time; it is constant, all the time.
Per: But that thing with showing off with a mobile phone, that was a short time. But it is pretty normal [to have a mobile phone], so that should go away. It is just a good thing to have a phone with you, you know.
Sigmund: How normal is it actually? I have heard that many people have gotten one for Christmas, but they have not gotten their bill yet. I wonder if there is going to be, you know . . .
Per: But there are more and more all the time. It is the same as other technical things. First come the show-offs and after a while it is OK.

These comments indicate that in 1995 there was not a settled agreement as to the nature of the device. The mobile phone was still undefined. The users had, perhaps, worked through their own adoption of the mobile phone and their own personal sense of the device. The comments of these respondents show, however, that there was no mutual understanding of the situation, and they were thus free to make their own attributions. One line of thought was that its use could be an indicator of suspicious people such as pushers and "real criminals." On the other hand, the last points of the same dialogue suggest a more normalized understanding of the device. A major element in the comments was that if one wishes to avoid being tagged as pretentious, or alternatively as a miscreant, then he or she should not use a mobile phone. Putting this into the terms of Silverstone and Haddon, the mobile phone was not completely domesticated (Haddon 2003). We were unsure as to others' interpretations of our use. The transition was on its way, however, since "after a while it is OK."

Another thing that augured against the adoption of mobile telephony was cost, and particularly, the cost of use. For some people, expense was not an issue. Early users with well-paying high-status jobs were able to afford a mobile phone, or more correctly, their companies were able to afford it. Indeed, it was often seen as a business expense. We saw this attitude in the swagger and braggadocio of Jakob, who was "on several boards" and talked of spending 32,000 kroner for his first mobile phone. Others were more careful when judging the expense of a mobile phone. The comments of Herald, interviewed in 1995, show the sensitivity of this issue:

> I actually don't want a phone that is more expensive than the normal [landline] telephone, normal subscription price like that. You almost need to have two telephones in a way. I am married and live with my wife. It is not necessary to have a mobile phone because then everybody has a telephone in a way. We have to have one in the house and so it would be an additional [one] regardless of how you look at it. So you buy two mobile phones [and] then it is almost a necessity that it is as cheap as a normal telephone.

The pricing of landline telephony was a salient price boundary. It is also interesting to note that Herald had started to frame his thoughts about the mobile phone in terms of its being a personal item. This is a paradigm shift when compared to the landline phone, which was often shared by the whole household.

"Don't Sit There and Blab into a Mobile Phone": Mobile Phones and Poor Etiquette

Beyond being a contested status marker, using a mobile phone in public was (and is) sometimes seen as boorish. The correct behavior when using a mobile phone is an area where we have had to work out our collective understanding. In addition, etiquette, and the breach of etiquette, allows us to measure a situation against the somewhat fixed notion of what is proper. It is clear that some blurriness surrounds our ideas of correct manners. As new elements come onto the scene (e.g., the use of mobile phones in public), they disturb our preexisting notions of propriety. Like Garfinkel's (1967) breaching experiments, these local disturbances give us a glimpse at the difference between the established and the newly arrived. At this early point, the affront to tradition can be used as a way to justify nonadoption. Steiner's comments that teens' use of the mobile phone was "all screwed up" illustrates this.

The mobile phone allows the abrupt entry of what Goffman (1963) would call "side engagements" into our daily lives. Turning again to the genre of urban myths, one story describes a mobile phone user who talked long and loudly on a bus. Finally, an exasperated listener grabs the phone and throws it out the window or smashes it on the floor. Another version of this story describes a retired police officer riding on a train who yells at another passenger for talking on his mobile phone and then strikes the hand of another female passenger who tries to intercede (Rosen 2008). In yet another example—captured on video—a so-called bus uncle in Hong Kong becomes unhinged when another passenger talks on his phone on the bus.[4] The mobile phone can clearly stir deep emotions when it comes to our behavior in public (Monk et al. 2004).

Often, when we are confronted by the mobile phone in public situations, we try to deal with it as best we can. Indeed, as the domestication approach would suggest, there is an emerging sense of where and how the mobile phone should be used. Almost from the time of the mobile phone's commercialization, the new and often unwelcome potential to interrupt others with its use was the cause of many comments. These comments were often made in the service of supporting the status quo. In the interviews from 1995, respondents talked about how phones were being used in restaurants and meetings (see also Ling 1997):

> One place where it should be turned off or put on voice mail is [in meetings] if there are twenty people sitting there, who sit there and call, and it rings, and if everybody had their own phone and they started to call, I don't think that is a place that you should have them. That is my opinion. Then you should set them over to the pager. That is in the meeting rules, you know . . . you can check it every now and then. It should not ring and make noise. I think it is wrong to use it there.

Teens interviewed in 2000 also commented on appropriate and inappropriate locations for using the mobile phone. They felt that it was not appropriate to use the mobile phone at the movies and other focused gatherings. Others cited examples of how solemn occasions were out of bounds:

> Kerstin: There are some simple Norwegian norms and normal courtesy. OK, our parents were not raised with the mobile phone, so they probably have not pounded the special rules into us, but these are things that you think out yourself. I got angry when I was at my grandfather's funeral and a mobile phone rang, you know. First, you are in a church, not that I am so Christian, you know, but you should have a little respect for the people who are there. There is a dead person there, you know. Then you don't sit there and blab into a mobile phone.

We have started to develop ways of dealing with mobile telephony in public. Texting is one of these adaptations since it does not interrupt copresent situations. In addition, when engaged in a somewhat longer call, we often seek out a quiet corner away from others so we don't disturb them. Finally, if we receive a brief call, others who are present have learned to redirect their attention away from the person using the phone for the duration of the call (Ling 2004).

This does not mean that these intrusions are accepted. During the initial popular adoption of the mobile phone, many people commented negatively on those who made and received calls in restaurants (Ling 1997). The use of the mobile phone while dining was seen as a breach of courtesy. Indeed, dining is one of those social situations that is perhaps most carefully governed by rules. More recent data from 2009 show that texting has perhaps replaced talking on the phone as a point of contention during meals. A teen girl, Carrie, said:

> my little cousin, she just got a Blackberry. . . . She sat up there, and she was texting at Thanksgiving last year, and my aunt's like "I'm going to break that phone." I mean she was like, "look at Carrie, she's not texting!" But still, . . . I honestly think it's really disrespectful, you're with your family.

Another teen girl said, "[My parents] only get mad if we're like eating dinner or if we're on it all the time and they're trying to talk to us. And they're like, um, put the phone down." The material shows that adults are also vulnerable to this critique. Another teen girl reported that "My mom only gets mad when we're out to dinner, because my dad always takes it out because he has a Blackberry and it's just glued to his head. Or his face."

Although the nature of use might have changed during the period since 1995, the discussion is still alive. The growth of the mobile internet gives us a new reason to divide our attention between the copresent and the remote. While we have some techniques for dealing with the mobile phone's intrusiveness, we are still in the process of determining how to use it in our copresent lives. One front in this process is the development of our collective sense of where, how, and how often it is to be used. The discussion regarding manners and courtesy is a discussion of the balance of attention between those who are copresent and the mediated connections who might call or text. The comments of the informants show that the narrative of the "mobile phone as an interruption" is still with us. Our sense that it is inappropriate to use the mobile phone in some social situations is—perhaps rightly—a dampening effect on its use.

"You Want Your Space, but When You Have Your Phone You Can't Have Your Space": The Stress of Constant Access

Another line of anti–mobile phone commentary is that it is stressful. The mobile phone allows others to call us when we would rather not be bothered. People often make this argument with an implied sigh, since they are already mobile phone users and see this as unwanted but unavoidable side effect of the mobile phone. Karin, a woman in the 1995 interviews, said:

> In some ways, you are never free, and if you have a job where you need that mobile telephone, sometimes I feel that we could have turned it off, but in relation to work he [her husband] feels that he needs to have it on. Either you are free or you are not. During Easter, we usually have 25 to 30 calls a day. Then you start to wonder if you can have a vacation or not. Of course, we have both, but there are some disadvantages. OK, it is his way of being responsible. But in other situations I see that it is positive, since we have older parents and we could not have gone away if we didn't have the [mobile] telephone and they can get in touch with us. So I see that it goes both ways. I see that we cannot avoid having it. We have had one since '82 or '83 and we are either abused or spoiled by it. It is difficult to say.

Karin's comments underscore the tension between being connected and being stressed by the mobile phone. She felt that having a mobile phone made it difficult to relax and get away from work. At the same time, she saw the benefits of having a mobile phone, as well.

Some early users tried to limit the stress by turning on the device only when they wanted to make a call. Others found other ways to limit access. Joachim, interviewed in 1995, said:

> I am not in the telephone book . . . My closest [friends and family] know that I have a mobile phone, and I know a lot of people, and my customers know that I have it and on my business card, but not in the telephone book. I believe in keeping my private life free with the mobile phone.

At a time when the mobile phone was reaching critical mass, Joachim was trying to strike a balance between limiting broad access to while allowing those people who were a part of his inner circle to reach him. He felt that once his phone number became publically known, others would impose themselves into his private sphere, an issue that had been seen as a problem with the introduction of the landline telephone almost a century earlier (Martin 1991).[5]

Going forward to 2009, the widespread diffusion of the mobile phone meant that the threshold for contacting one another had been lowered dramatically. While caller ID meant that we could screen unwanted calls, there was the growing expectation that we should be available via the mobile phone. For teens, this means that their parents can call and remind them of arrangements and perhaps invade their social interactions. One teen girl, Blake, interviewed in the United States, said: "The worst thing [about the mobile phone] is that you don't want to get in touch with your mom but she can always get in touch with you. You want your space, but when you have your phone you can't have your space." In the case of Tom, a teen from the United States interviewed in 2009, the mobile phone was a channel through which classmates could pester him: "There are people that are like one or two people and it is like 'Please don't call me. Please don't call me.' And they are constantly [calling]."

The comments of Blake and Tom illustrate the way that power relationships are being exercised through the mobile phone. In the case of Blake, the mobile phone, at least partially, is not a device of self-realization, but something that approaches a device of subservience. Her mother has more direct control than Blake would wish. Indeed, for many women in developed countries (Rakow 1988; Rakow and Navarrow 1993; Chesley 2005) and in developing countries (Wallis 2011), the mobile phone is a device that can reinforce power differences. It is a way for people to receive the commands and requests of others. While we can cultivate identity development via the mobile phone, it can also confirm our subservience (Elliott and Urry 2010).

The comments of the informants do not necessarily imply that it is bad to have a mobile phone; rather, they suggest that it needs to be used with caution. The stress of being constantly available and the fact that the mobile phone can make power relationships uncomfortably explicit are justifications for not having a phone, or for using it judiciously.

The Mysteries of the Technology

One final line of discussion arguing against the use of mobile telephony is the ability of its radio signals to generate untoward side effects. Some of these are the object of serious research, for example, the effect of electromagnetic energy on users or the loss of honeybees because of the placement of cell towers. In these cases, serious and indeed impassioned researchers are investigating the different dimensions of these issues.[6] In other cases, the interaction may be more spurious. Interviewees in the 1995 focus groups, for example, reported that mobile phones had set off alarms and locked car doors, among other sorts of mischief:

Erik: It [mobile telephony] is misused in a big way. For example, school children have started to take it to school. It rings in the classroom. The fire alarm goes off because it . . .

Oddvar: It is like you said; it set off the fire alarm . . .

Arne: I have heard people talking about explosions [at construction sites] also.

Dag: It is prohibited at the train station in Stockholm. You cannot use a mobile phone at all the big hospitals.

Erik: I read that it had, as a matter of fact, affected the traffic regulation and traffic lights.

Dag: That is the new GSM.

Oddvar: I heard that people said that they were not allowed to use the mobile phone, but there were so many business people who sat there and talked, and they said that it was a completely new telephone, a whole other system, they didn't react to the alarm . . . but in that building it is forbidden to use it since it sets off the fire alarm.

Arne: I had my mobile phone sitting in the car, and someone called me and I was not in the car, and I have a central locking system, and because of that I was locked out.

Other respondents added to the list of problems with the mobile phone. It was suggested that it disturbed the gas pumps at gas stations or even worse ignited the gas, causing an explosion.[7] These are all examples of contemporary folklore associated with the unknown effects of the mobile phone. It has been claimed, for example, that the electromagnetic radiation from cell phones can pop popcorn or cook eggs.[8]

Such assertions of incomprehensible side effects from technology are not new. Caroline Marvin has described how the introduction of electricity into the home gave rise to the idea that lightning was more common, since electricity was "leaking" from the wires, just as gas leaked from gas pipes (Marvin 1988). These stories tell us about our relationship to technical developments in society. The threat from technology comes from its mysterious workings. These contemporary legends are often founded in the clash between more traditional lifestyles and modern developments (Best and Horiuchi 1985). They are not related to supernatural individuals as in traditional myths; rather, they are grounded in human action and our collision with new technology.

According to Brundvand (1981), urban myths are particularly important if they involve threats to society (particularly children), fear of crime, or physical danger and mistrust of others. Examples include the partially decomposed mouse in the soft drink bottle, the spider in the woman's hairdo, and the fried rat. Gary Fine (1973) writes that these stories arise from the insecurity associated with broad transitions to society. Specifically, they arise with the rationalization of society, where we substitute formal relationships for what had been informal interactions. Urbanization, privatization, and value shifts are often the basis for attributing unpleasant effects to technical developments (Best and Horiuchi 1985; Fine 1979). Changes to the structure of society and disaffection in daily life provide the basis for urban legends. The legends provide us with a way to think about the impact of, for example, technologies on society. So long as there is a certain potential level of credibility in the legends' assertions, they give us a way of illustrating how powerful or destructive technologies might be.

They can also feed into a broader ideology opposing the adoption of the technology. In other cases, they give us an account outlining how we are being misled by large corporations, the government, or other omnipotent forces. In addition, since urban legends always recount a second-hand experience, we have the ability to distance ourselves from such stories. Should the assertion be unmasked as a fake, we have the opportunity to say it was just something that we had heard or read.

Urban legends are a way to help us understand how people frame the dark sides of mobile communication. The rise and persistence of this form of contemporary folklore indicates that we are not completely comfortable with the values implied by the diffusion of technologies into society. The various assertions have a certain tenacity. The doggedness of these tales may just as well be a marker of our broader concern with the social impact of the mobile phone. The use of these devices in new circumstances gives rise to assertions that they are somehow a threat or can present a risk.

In summary, several prominent story lines, for a variety of reasons, urge caution with the adoption of mobile telephony. As mobile telephony has approached critical mass, people have constructed a series of narratives that suggest that the device should be avoided or treated with caution. These naratives have changed as mobile telephony has become more common. The narratives started with the idea that mobile telephony should be reserved for serious use, that is, for business. As this bastion fell, the idea arose that mobile phone use was a breach of manners, stressful, expensive, and that mobile telephony could have unanticipated negative effects. Each of these points has been argued with differing intensity and has been sustained or discounted. We have used these narratives to explain the device and perhaps to justify nonadoption (Berger and Luckmann 1967). The diverse elements—showing off, breaching etiquette, expense, and stress—associated with the mobile phone were gradually woven into a more general ideology opposing the mobile phone (Flichy 2007, 11). Ultimately, however, the weight of these arguments has not hindered the widespread adoption of mobile telephony. Indeed, the positive side of mobile communication also has its legitimations.

"It Really Makes You Popular": Positive Accounts of the Mobile Phone

The list of reasons not to have a mobile phone is long. Taken at face value, these counterjustifications might be enough to ensure that no one would use a mobile phone, or if anyone dared to, he or she would use it discretely so as to not be observed. This is obviously not the case. Rather, people have adopted the mobile phone in ever greater numbers. Just as there are narrations against the use of the mobile phone, there are those that support it. It is these accounts—sometimes commercially supported—that allow us to construct plausible justifications for buying and using one (Berger and Luckmann 1967). The construction of legitimations helps to affirm "how things are done" and allows us link the adoption of the device to other broader social values such as security, sociation, and accessibility.

We have a constantly changing iconography of mobile phones that are popular at any given time, and we perhaps covet the newest, sleekest models. The cutting edge of mobile phone fashion has moved from, for example, the Motorola Dynatac, to various Nokia models, to the Motorola Razr, to the iPhone, and eventually to another device du jour. As we are enticed by one or another device, we feel that the mobile phone makes the world more secure, giving us immediate access to one another and more control over the flux of daily life. We hear about how it supports economic development and simplifies the lives of people in the global south; we take heart when we hear that the mobile phone has played an essential role in an emergency. We allow ourselves to be beguiled. We are convinced that it not just "nice to have"; rather, it is essential. This willingness to believe in the potential of the device facilitates its integration into our daily lives and facilitates the sense that the mobile phone is taken for granted.

"If Something Should Happen": Security as a Justification for Mobile Telephony

As is well known, mobile phones have gained a central role in dealing with urgent situations. In the next chapter, I will look at how it has restructured our response to emergencies. In addition to making these structural changes, the sense of security provided by the mobile phone is often used as a legitimate justification for owning one.

The mobile phone has been used to facilitate responses to various emergencies and to check on the status of loved ones who may have been in harm's way (Cohen and Lemish 2005; Dutton 2003; Katz and Rice 2002). Adventurers who have lost their bearings or who have met with unexpected situations have used the mobile phone to get them out of their fix. Traffic accidents, fires, heart attacks, and other calamities both large and small are regularly reported via the mobile phone. As we will discuss in more detail in the next chapter, the mobile phone has changed how we deal with safety issues and how we respond to emergencies. In addition, its use in these situations has helped us to legitimize the device.

Safety has been generalized into a justification for always having a mobile phone. It is a common refrain among interviewees who have a mobile phone in case "something happens." This generalized justification is the result of individual events and instances that support a broader ideology. These events include things that we have experienced ourselves, but also the stories of other people. The broader ideology can be seen in the exemplification of Sid, a teen interviewed in the United States in 2009:

Sid: I leave my phone on all night because I get paranoid, like if someone needs to call me for like . . .

Interviewer: Something bad happened?

Sid: Like . . . A few years ago I was in California . . . I turned my phone off at night, and that's why I don't do this anymore, I turned it off and then my cousin's buddy, who's like also friends with me, was driving through the mountains. It's basically like here, they're curvy and windy, [he] flew off the side of the road and he was pinned up against a tree and he kept trying to call 911 but for some reason it wouldn't connect to 911, so he called me. And he kept calling me, and calling me, and calling me, and my phone was off, and he died. So if I had my phone on, you never know.

Sid's story presses, and perhaps exceeds, the limits of credulity. In spite of this, it is a wonderfully condensed narration of mobile-phone-as-safety-link. The moral of Sid's story is that we should be available via the mobile phone for friends who are in need. It is almost possible to hear Sid telling this story to a teacher or parent who wants him to turn off his phone and, for example, concentrate on his homework. In Sid's mind, his story justifies being constantly accessible.

This desire for safety and security has become one of the ways that we justify ownership of the device. We keep the phone with us at all times as a precaution. Some of us may never need the mobile phone to report an accident or to get out of a dangerous situation. Nonetheless, we have the mobile phone with us as a sort of insurance (Cohen and Lemish 2005).

Elderly respondents in 1995 said that they bought the mobile phone so that they could go skiing or be out in nature and still be able to call for help should the need arise. One respondent, from the 1995 interviews, Dagny, used this justification when she said:

> I bought it simply because I go to the mountains alone. Suddenly I get car trouble in –20 degree weather. What am I going to do when I am sitting there? That was the main reason I bought the mobile phone. Plus, I have a cabin without electricity. What if it starts to burn when I am alone? What am I going to do then? It is security when you travel alone.

There is certainly some fuzzy reasoning underlying these comments. Accidents can happen in spite of having a mobile phone. Using the mobile phone in rural areas can be dodgy since the signal might not reach a base station. The eventual location of a skiing accident may be deep in the forest and not accessible by helicopter; weather conditions might not allow for rescue, or the battery might lose its charge in a cabin with no electricity. Regardless, both Dagny and Sid had constructed justifications for having a mobile phone based on notions of safety and security. They both legitimized their use by seeing it as a safety link.

Beyond the evidence provided by the group interviews, a 2009 analysis in the United States showed that 94 percent of parents and 93 percent of teens aged 12 to 17 agree with the statement "I feel safer because I can always use my cell phone to get help" (Lenhart et al. 2010). Indeed, the analysis shows that 100 percent of girls aged 12 to 13 agreed with this statement.[9] These findings, along with the material from the interviews, show that we see the mobile phone as a security link. It gives us the freedom to extend our boundaries. It allows us to go skiing or to be at the cabin on our own. It gives us a sense of security when walking late at night in dicey parts of town (Baron and Ling 2007). It makes us feel safer to know that the mobile phone will be there should something happen and the need to summon help arises.

The sense that the mobile phone is a safety link has also been operationalized in other ways. In the words of Joanne, a Norwegian teen interviewed in 2000: "If I am walking at four in the morning, it is very nice to talk with someone on the mobile phone as you are walking down the street. You are safe because if something happens then you are in direct contact with someone." Using the phone to talk with someone—or in some cases to pretend that you are talking with someone—plays on the idea that being in touch can ward off potential problems (Baron and Ling 2007). The teen girl cited here talked about actually being in contact with another person at that early hour. Other teens described simply faking a call to deter attackers.[10] The fact that this practice has arisen shows yet another dimension of the mobile phone as a safety link.

Early adopters clearly relied at least partially on the idea of security in their justifications for having a mobile phone.[11] As mobile phones diffused into society and teens started to have mobile phones, the sense emerged that they provided a link between parent and child. Teens in the United States in 2009 said that keeping in touch with parents was one of the justifications for getting a mobile phone. One teen girl said:

> I think I got mine in fifth grade, and it was mainly because my parents wanted me and my sister to have one like after school to call them, and make sure we were safe, it was pretty much for safety.

Teen boys gave the same reasons:

Steve: [I got it] because my mom wanted me to call her, all the time . . .
Andy: I got mine when I was about twelve, and I got mine to keep in touch with my parents . . .
Jack: I got mine when I was eleven.
Interviewer: Same reasons did you say?
Jack: For emergencies.

The mobile phone can be said to facilitate the emancipation process, since it gives teens freedom while also providing them with a link to their parents should the need arise, although to be sure, there is debate over whether this helps or hinders the teen's process of growing up (more on this in chapter 7).

While teens accept the pestering of their parents in return for the ability to have their own mobile phone, parents offer another story line, namely, that it provides safety as well as a latent communication channel. Parents say that the child can expand his or her radius, and gather some of life's ballast while still operating in a relatively safe haven. Purchase of a phone is sometimes prompted by a situation where the child is caught in some minor emergency, such as missing the last bus on a rainy evening. The mobile phone is seen as a bulwark against the uncertainty of children moving beyond the sphere of the home. In addition, it is seen by adults as a way to contact others "just in case something happens."

Access and Autonomy among Teens

The mobile phone has become central to the lives of teens. They see it as a sign of independence. More than anything else, according to teen respondents in the US focus groups in 2009, the mobile phone was the way to contact friends (Lenhart et al. 2010). Indeed, texting has been found to be the most common way for teens in the United States to contact one another. "I can talk to people whenever I please," according to a teen interviewed in the United States in 2009. Another teen, Taylor, interviewed at the same time, said, "[It helps you] staying in contact with everybody." It was seen, perhaps unreasonably, as the gateway to popularity:

Lyle: I think the best thing about phones is when you get a lot of kids' numbers it really makes you popular, and even when people you don't really know text you, it just kind of shows you how everyone knows you.

The mobile phone has become the locus for access. If it is lost or broken, teens lose more than the physical handset; they also lose access to their friends. In the words of Jaisa, interviewed in the United States in 2009, "If it breaks or something, you'll, you've lost your numbers." The primary justification for having a mobile phone is access to friends.

In addition, the mobile phone is an important personal artifact. Shandra, a teen from the 2009 group interviews in the US said, "it is yours. It is your own personal thing. You can decide on the ring tone and on the color and like it is yours!" Another US teen, Sarah, said, "It makes you feel older and independent, I think. I think that is what it comes down to." This independence comes into play as teens move toward emancipating themselves:

Nandi: The best thing is that my mom, like, she can't get on the phone and be like, who are you talking to, like, in all of my conversations. And I can go to my room and have, like, a private conversation.
Amber: Um, I think that the best is that you don't have to worry about anybody else butting into your business. 'Cause it's yours and you can, like, choose what people can and can't see. And you have, and it's like your own privacy.

Access to friends, autonomy, convenience, and personal control of the communication channel are the lines of these narratives. The teens cited here have a special reading of these issues. They are in the process of working out their adult relationship with their parents, and they are also working out their friendships and relationships with eventual partners. The mobile phone makes obvious these different aspects of their lives. It makes them take a position in relation to always answering the phone when a parent calls, and it causes them to think about how and when to call friends when there is always a direct channel available.

"I Left without the Phone One Time . . . I Felt Completely Lost": Controlling the Flux of Quotidian Life

The justifications for having a mobile phone become especially obvious when we do not have one. During a series of interviews among Norwegian parents in 2005, we asked the simple question: "Can you imagine a day without a mobile phone?" Many parents thought that it was possible. Others were not so sure. One mother said that she would feel "naked." She continued, "It has happened that I have forgotten my mobile. It is like forgetting a purse. It is that feeling . . . You don't do it." Another father said, "You get a little panicked. I am very dependent on it and I forget it often and then I get panicked." The mobile phone helps us plan activities and make changes as competing demands arise. As with the sense of security provided by the phone, we have the sense that it helps us to manage the flux of daily life. We describe how it helps us to manage work and arrange activities with partners, children, and friends. In short, it gives us a sense of control in our daily lives. It has become a common tool for juggling responsibilities at home, at work, and in the social sphere. We have seemingly accepted the idea that we need to have it with us at all times. Not having our mobile phone prompts us to take stock of the situations where we might miss out. This is seen in the comments by participants in the 2009 focus group in Norway:

Jeni: I left without the phone one time. I got to work and I felt completely lost.

Interviewer: What was the problem?

Jeni: You know, communication with the world.

Interviewer: Did it have any consequences?

Jeni: Yeah, there were some meetings. We have a lot of floors in the building and we send messages to say that you are going to lunch or that a meeting has been postponed. We can send text messages to telephones from our PCs. When I didn't have a telephone, it was almost so bad that I had to go home to get it.

Interviewer: Did you miss anything?

Jeni: No, I didn't, but I could have. That a meeting was moved, or something. I have old grandparents who are not in the best of health. As a nurse, I am afraid that something might happen.

Interviewer: You are afraid that something would happen and they could not get in touch with you.

Jeni: Yeah, the landline phone is like not in use.

John: I have done that [forgotten a phone].

Interviewer: How did the day go?

John: It wasn't good, you know. I had to go home and get it.

Nothing went wrong for Jeni on the day she described, but that did not assuage her sense that it might have. Both she and John express the idea that we need the mobile phone with us at all times.

As we have the sense that the mobile phone has reached critical mass, not having one with us means that we are unable to deal with a whole set of demands that might arise. Jeni knew that planning took place via the mobile phone and that it was the link to her grandparents. Forgetting her phone meant that she was out of the loop. At some level, it meant that she was not being a responsible coworker or family member. She felt that she might miss unannounced events and that she did not have access to telephone numbers or appointments. She might miss a text message with the address of a restaurant where she will meet a friend, or the like. This is not to mention the flow of texts and short calls that she might use to touch base with friends and colleagues and to monitor what was happening with family and friends. Like a purse or billfold and keys, we feel we need to have it with us. If not, we are somehow not a fully participating member of society.[12]

When we forget our mobile phone we find ourselves fumbling in the dark, or at least in the twilight. Just as with the justifications opposing the mobile phone, the positive accounts are woven into an ideology that helps us understand why having a mobile phone is important in the ongoing production of our daily lives. These justifications also tie into broader social themes of safety, efficient coordination, and concern for our loved ones.

Conclusion

In the early 1990s the marginal position of the mobile phone meant that it was not often the focus of negative commentary. As it became popular, however, two broad narratives arose. On the one hand, there were discussions of how it was a disruption to our lives, and on the other hand, it was seen as helping us to order our affairs. The latter storyline has shown itself to be the most powerful. Put into the context of Giddens's notion of ontological security, since the 1990s we have seen the evolution from one regime of the traditional order based on landline telephony, through a transitional period, into a second regime that includes the mobile phone (Giddens 1981, 82).

Following the line from disruption through the transition into a new regime, we can see how the various discussions of the mobile phone have played out. In the era of the landline telephone, we had an established sense of where and how to hold telephone conversations. They occurred in a specific place—in the hallway at home (Gittu, Jørgensen, and Nørve 1985), in the office, or in a telephone booth. The adoption of the mobile phone challenged this established practice. We then developed a set of practices and justifications surrounding the new arrangement. When business people started to adopt the mobile phone, this caused no major disruption to our sense of telephony. The technology's diffusion was somewhat limited and novel, and there was a readily available justification associated with its use.

As costs came down and mobile phones spread to other sectors of society, they were often seen more as an affront to our sense of public decorum. Their use was not in the service of important decisions and the command of the economy, but rather for the coordination of mundane activities and the exchange of gossip. This rubbed many of us the wrong way. Using a mobile phone was no longer a sign of status, but of vulgarity. It was not a sign of efficient management, but of profligate decadence. In some cases, as noted above, use of the mobile phone was seen as a threat to public safety. This storyline has described how the mobile phone gives rise to status disruption, bad manners, wasting money, and stress. These discussions have helped to advise us on the way that we use the mobile phone by helping with the adoption of mobile phone etiquette and by delimiting when and where we are available to our different circles of friends and colleagues.

The alternative account has focused on the positive sides of mobile telephony. As time has gone on, we have established a sense that the mobile phone provides security and that it assists us in the completion of our daily tasks. Indeed, we have begun to feel a lack of security if we forget our mobile phone. We might be missing important events, or there might be people who need to get in touch with us.

Thus, there have been three broad and overlapping phases in our understanding of the mobile phone. In its early life, the mobile phone buttressed the status of users by differentiating them from others. As it was popularized, we experienced discussions of status disruption, etiquette, and a sense that the device was a threat to public safety. Simultaneously, the mobile phone became a common way to organize the many details of our lives. Finally, many people have come to see it as a great inconvenience, if not a catastrophe, when we forget or lose our device.

Following the notions of Berger and Luckmann, we have seen the clash between the anti– and the pro–mobile phone positions being worked out. Each perspective has drawn on different value structures when working out its position (Berger and Luckmann 1967, 112–113). Ideas of status disruption, cost, or perhaps breach of etiquette were drawn upon when constructing a "subjectively plausible" ideological structure opposing mobile telephony (ibid., 110). On the other hand, notions of availability, safety/security, and coordination (not to mention a fascination with gadgets) supported the adoption of mobile phones. It is also clear that there were and are significant commercial interests that also support the latter justification structure while trying to undercut the former (ibid., 127).

In the past two decades, we have been working out a type of accepted or correct behavior vis-à-vis the mobile phone. As we move toward an understanding of the mobile phone, we encrust our attitudes with a type of moral authority. This gives rise to thoughts such as "People who use a mobile phone in a restaurant are impolite," or "It is irresponsible to forget your mobile phone when you are downtown." People who do these things are not playing by the rules.

In addition to a transition in our attitudes toward the mobile phone, we have also seen it establish itself and rearrange the social ecology of communication. The diffusion of the device into society and its legitimation have led to a society where we have changed how we coordinate our activities, how teens grow up, and how we keep ourselves safe and secure. It is to this issue of safety that we turn in the next chapter.

7 Mobile Communication and Its Readjustment of the Social Ecology

The person-to-person accessibility afforded by the mobile phone has introduced a new logic of coordination, safety, and family organization, as well as a structure for the way that we arrange our work, schooling, leisure time, and political activities (Urry 2007, 179). In this process, it has outcompeted other forms of mediated communication.[1] It has gone from being a marginal artifact to being the most widely diffused technology of social mediation. As we have seen in the previous chapter, it started out as a tool for business and emergency services, and it has grown into a *de rigueur* part of our kit, through which we contact one another and arrange our lives.

Following Boulding's (1978) notion of a niche, this large-scale adoption of the mobile phone continues to change the communications ecosystem. Just as the rabbits in Australia moved into an open niche and then rearranged the conditions of the contiguous niches, the mobile phone has rearranged the conditions of our social interactions. It has changed the communications niche by pushing aside those technologies that previously occupied its space. The adoption of the mobile phone has reduced the market for landline telephony and has seen the near extinction of traditional telephone booths (Ling and Donner 2009).[2]

As mobile telephony moves into this space, it is outcompeting alternatives such as landline telephony and public phone booths. According to material from the ITU (2010), over 100 countries reported fewer landline subscriptions in 2009 than they had in 2004, and another 22 countries had only marginal increases in the number of landline phones. In many developing countries, the landline system was at best wobbly to start with, and the rise of mobile communication has undercut that already unstable base. During this period in Ghana, subscriptions to the landline system fell by 13 percent, while the mobile system grew by 70 percent. In 2008, 85 percent of the all telephone subscriptions in Ghana were mobile. In

Bulgaria, the landline system fell by 4 percent, while the mobile system grew by 24 percent. This phenomenon is not only found in the developing world. In Denmark, subscriptions to the landline system fell by about 7 percent. There was a nearly symmetrical increase in mobile subscriptions.[3] The same was true of Finland (–9% landline and +7% mobile), Italy (–4% and +10%), Austria (–3% and +8%) and the United States (–3% and +11%). On a global basis, the number of landline subscriptions grew by 2.4 percent while the number of mobile subscriptions grew by 24.3 percent. All of these numbers show that the mobile phone is moving into territory that was formerly occupied by the landline telephone and changing the landscape of telecommunication.

To be sure, mobile telephony built on the foundation of the landline phone, which had brought telecommunication into the home. The mobile phone has individualized communication and has given us the ability to interlace communication with other ongoing activities. It is also increasingly giving us access to information via smartphones and the mobile internet. As a result, we are becoming more thoroughly enmeshed in our personal networks (Licoppe 2004; Wellman et al. 2005). The mobile phone has restructured how we coordinate our lives, keep ourselves safe, and organize our family and social lives. The mobile internet and smartphones are adding new dimensions to these trends. We can easily make and break appointments, and we have access to a wide range of information, which we can pass on to others more quickly than ever before. The mobile phone has restructured the way that parents keep tabs on their children and has given teens a new freedom and independence. It has changed how commerce takes place and how we participate in the public sphere. In short, mobile communication has changed the ecology of communication.

"A Little Back and Forth": The Changing Method of Coordination

Microcoordination and Social Networking Mobile Memes

The mobile phone has altered the underlying structure of how we coordinate activities. We are always individually available to one another, and we expect others to be available when we call. We are always—at least theoretically—a node in the social network. Indeed, it is sometimes more remarkable when we are not mutually available.

As I have noted, clock time is a central coordination tool for society. However, mobile communication, with its focus on individual availability, is supplementing, or perhaps modifying, time as a coordination tool.

The ability to microcoordinate is new, but it has not replaced the need for clock-based coordination, particularly in larger, formal organizations. For teens on weekends and for families at the shopping mall, microcoordination works well. For people with tight formal schedules, the ad hoc nature of mobile-based microcoordination does not work.

In those situations where microcoordination is possible, however, mobile communication allows us direct access to the other person. We often do not even need to agree on a particular time and place to meet. Rather, we can iteratively work out the time and the place to meet in a series of calls and texts (Ling 2002).[4] The ability to be individually addressable is a minor paradigm shift. It means that the former regime of agreeing on a strict time and place for meetings is no longer needed in many cases. This is particularly useful for teens, who are not as deeply ingrained in traditional forms of time-based coordination. Agreements may be couched in a vaguely agreed-upon temporal structure that has been progressively refined through a series of interactions. Hege, a female in the 2003 group interviews, describes this process:

Interviewer: How do you decide on how you are going to go to the movies, how is it done?

Hege: I have an agreement with my friends to go to the movies in the weekend, then we have agreed Friday or Saturday, not more specific than that, but we both have a mobile so we can reach one another whenever we want and agree later.

Interviewer: Do you call or send a text message?

Hege: Then I call.

Interviewer: And when Friday comes . . .

Hege: When Friday comes, I call.

Interviewer: "We had an agreement what do you want to do . . ."

Hege: "Should we meet in the evening or in the morning?"

Interviewer: So it gets specified.

Hege: Yeah, and then we text a little back and forth.

Hege describes the iterative process whereby she and her friends start with a vague notion that they will meet up together. As the situation develops, they have a better sense of when they might be available and what activity might be of interest. This openness, however, means that there may be several agreements "on hold," and the individual may prioritize and reprioritize things as the situation develops. At one moment, it might be that a movie with their friend Carlos is of interest, Tanya might ask if you are available for dinner, and Cody might suggest that there is a party the same evening. Since none of the agreements is fixed, each of them rides until a final decision is made.

Participants in a 2003 group interview were divided on this type of situation. Some teens thought it was a reasonable way to work out their social lives, while others found it stressful. For some, there were too many moving parts. There were too many times and places to work out and too many judgments to make. It was difficult to manage all the different possibilities. One interviewee, Per, said, "There is more stress when you have several things going on at the same time." It can mean that we have to reply to different people with various versions of the truth, or it might mean that we need to engage in complex scheduling so as to maintain several facades (Ling 2004). It can be stressful having to manage several different potential encounters at the same time via the mobile phone. Some had developed techniques for managing these situations. The mobile phone was a way for them to extract themselves from the problem.

Martin: So that you don't get in a jam, then you can call and say, "I can't make it, it doesn't work out. If I am going to make it, you need to, give me an hour."

The adoption of smartphones and access to the mobile internet is also changing the way that coordination (as well as mobile interaction) takes place.[5] People are using smartphones to access social networking sites, Facebook in particular in order to coordinate activities and to keep up with the goings-on in their social sphere. Other sites, such as Twitter, Google+, LinkedIn, foursquare, TalkBox, WhatsApp, Skype, and Viber, were sometimes used, according to focus group informants, but Facebook was the most commonly used social networking site.[6]

Mobile Facebook facilitates group coordination. The "available" function in mobile Facebook allows users to see if their intended interlocutor is logged on. If so, they may be willing to open a chat. Another advantage to mobile Facebook (also found in PC-based access) is that the thread of messages is visible to all the participants. With texting, the person organizing, for example, a meet up at a cafe, might send a text to four people. None of the people who have received texts can see the responses of the other three recipients, only their dialogue with the person who originated the message, who serves as a "hub" for the communications. There is no way for the separate people to follow the comments of all the others. If one of the people receiving the text can suggest a better location, it means that the hub has to send out a whole new set of texts to the others in order to field the suggestion. By contrast, in Facebook, all the people included in the thread can follow the interaction, making it easier to organize things when several people are involved. There is no central hub. Anna said, "If you are going to meet five people for group work you will not send out a message to five people saying 'Where are you?' You just post on Facebook and check it on your mobile to find out where they are. It is easier." Another respondent, Pernilla, noted:

> I use Facebook if I need to connect to more than one friend. If it is just one friend I will always send a text message. If it is just two friends I write a thread on Facebook mail. I know that people check it almost as often as they check texts. You get an overview on Facebook because you can see what other people have replied and you can't see that in a text message.

The groups who use this type of coordination may be one-off, ad hoc groups (e.g., an assortment of people planning an evening out), or more stable groups based on either social or institutional connections. Some teens have reported that they have stable Facebook groups, such as all the boys or girls in class, a circle of three or four close friends, members of a sports club, or the students who have a common semester project. These groups can be used as needed for planning or for simple status updates. The content can come from PC-based access to Facebook, or it might include photos and text produced on the spot with a smartphone.

A problem with Facebook, and in particular mobile Facebook, is that not all people in a group necessarily use it. According to the participants in the focus group, it was not always possible to rely on the fact that everyone used mobile Facebook. In that case, texting, which is broadly diffused and thus more taken for granted, comes into play as the more reliable alternative.

> Mari: With my two best friends, one of them takes the initiative and sends text messages. And they respond back, and I respond back and say, "well, I can't do it on this day and she can't do it on that day. Can you do Thursday?" Then who ever initiates the contact is responsible.
>
> Stine: But it becomes more difficult if there are five.
>
> Mari: But they really don't use Facebook a lot . . . So we use texting a lot more.

Another interviewee, John, also noted that "It is the expectations [of the group]. With my group last semester, we just texted. They were more 'offline' type of people. It wasn't legitimate to use Facebook in that situation." Thus, for some groups, Facebook has appeal as a planning tool. If all the group members use it, it gives them a better functionality. However, some people do not have access to the mobile internet via advanced mobile phones, or perhaps choose to cling to texting for other reasons. This means that Facebook has a secondary position vis-à-vis texting or voice interaction when coordinating activities.

Mobile Facebook expands the functionality of group coordination. According to these participants, it was also used to keep a more abstract overview of the happenings in their various social groups. Using Facebook on a mobile device meant that they could follow the meme of the group.[7] The participants could do a quick check on the latest updates of their friends and acquaintances even when a PC was not available. The fact that we can take a photo and upload it directly in situ gives a multimedia dimension to these updates. Pernilla noted: "You can check what is happening more often. I can be in my bed or on the train and I can check on Facebook. It is not that I post more or write more, it is when I am bored that I use it for entertainment."

The posts on mobile Facebook can be more tied into the flow of daily life, and, as with the PC-based posts, they can be calibrated for the quasi-broadcast nature of Facebook. According to David, "I put [posts] on someone's wall but it is [also] meant for other people. Then they can comment on it. The group interaction is the important part. I put it on someone's wall, but it is meant for the friends we have in common. Not only on the mobile, but also on the PC." Postings can be greetings, photos, and observations, as well as notices of the large and small events of daily life, such as "getting a new job or buying a new house." Thus, beyond the functional use of coordinating activities, it gives people a venue where they can follow the ebb and flow of activity in their social sphere. This can tip over into being simply diversion or "entertainment" when nothing else is happening. One interviewee said, "I only use [Facebook on the mobile] when I am bored."

The interviewees, however, were not in agreement when it came to interactions on Facebook as compared to texting. One participant, Mari, said: "I only communicate with my weak ties on Facebook. It is really really rare that I will write to my best friend." Others disagreed. David noted, "You are not equally tied to everybody on Facebook. There are five or six that I am tied to. The rest are an audience. Facebook is not weak ties. I communicate with my strong ties." Thus, for some, Facebook is not as tied to the intimate sphere, while for others, the weak ties are only peripherally in the picture. In contrast to this, texting is for closer friends. Friends (in the traditional sense) received one-on-one greetings such as texts or calls, while acquaintances on Facebook were not treated as cordially. The closest friends, for example, would receive a "happy birthday" text message and perhaps a call. Acquaintances would get a more impersonal Facebook version.

Other issues also render mobile Facebook a rather blunt social tool. One of these is the quirky "push" function. Push describes the way that, for example, text messages are sent to the mobile handset. We do not have to log on or check in order to receive a text message. With texting, the audible tone or the flashing of the icon indicating a new text might be cause for a slight bit of excitement. If configured correctly, mobile Facebook has the same functionality. However, it can be quirky to set up, and it does not carry the same meaning for the interviewees. According to Tom, "You get notices of every little thing [from all your Facebook friends]. It is not just emails. It is like someone liked the same thing as you did, that someone posted something on one of your groups. And suddenly you have 20 in a day and it doesn't really count for anything." His sense was that the incoming information from mobile Facebook was not as individually pertinent and, in fact, could become a bother. Beyond this, mobile Facebook was not seen to be as functional as the PC version. The interface was more awkward, the small screen was problematic, and it was not as easy to type the messages.

Thus, social networking, and in particular Facebook, has gained a position in the mobile world. The fact that it can be followed from either a PC or a mobile phone makes it extremely flexible. In addition, the ability to follow threads means that group discussions are easy to carry out. Finally, it gives users a chance to follow the happenings of their social sphere. That said, not all mobile users have access to, or choose to use, mobile Facebook. In addition, texting and calling are often seen as more personal and authentic. The messages are created with a single person in mind and are not crafted for a broad audience. The personal nature of texts gives them more weight in the estimation of their social significance.

These variations of planning also show us that the individual addressability provided by the mobile phone has challenged, and in some cases subverted, the regime of time/place-based planning. The logic of microcoordination allows us to work out our social interactions in a new way. The adoption of advanced smartphones and access to the mobile internet also adds new functionality. Microcoordination is clearly not applicable in all situations. Nevertheless, mobile telephony has given us a new tool for social coordination, and we are working out how to best use it for the nuanced coordination of our social lives.

"It Gives Us a Lot More Freedom": Coordination in the Business World

It is clear that microcoordination is not only found in the private sphere. Indeed, the coordination function of the mobile phone was quickly recognized by early business users. As we saw in chapter 5, in the mid-1970s the vast majority of people who had a mobile phone in Norway were businesspeople (Johannesen 1981). This was the case for the greater part of the next two decades. Respondents interviewed in Norway in 1995 commented that the mobile phone provided logistical support for businesspeople who worked outside the office. It gave them more flexible contact with customers. One user noted, "At work, I could not have managed without a mobile phone. Completely impossible. I could not cover as large an area." The mobile phone allowed him to be more responsive to the needs of his customers while staying in contact with the other people in his organization. Another user, Oskar, said:

> It is clear that with sales—I work in both Norway and Sweden and a little in Denmark—and it is clear that without the mobile telephone we would not have the possibility to reach all the things we need to do. There would have to be more people involved. It gives us a lot more freedom. You are always in contact with the main office, and at the same time, you are in contact with the international suppliers.

Users were also happy with the ability to reschedule meetings and the ability to work out their daily rounds more sensibly. As noted above, participants in the 1995 interviews were often executives or upper-echelon employees who had their phone paid for by their job. In a relatively short time however, work-related coordination of activities had reached other portions of the workforce. As others started to use the mobile phone, they also started to use the same techniques for coordinating their affairs. This development was not just useful for the individual; it had become a part of the way that employees were available to their employers. It was a different and more individualized logic of coordination.

As the mobile phone became more common, employers used them to organize shift work, and employees who were "on call" used them to stay in touch.[8] By 2003, hospitals and restaurants were using the mobile phone as a way to call employees to cover different shifts. This can be seen in the description of Inga, an adult female respondent in the 2003 Norwegian interviews. Inga was a nurse working at one of the large hospitals in Oslo. She described the way that the mobile phone was used for the staffing of the different shifts at the hospital.

> Inga: [I got it] three years ago, when it started to come as a fringe benefit at work. Then it was a "sneak" benefit, a fringe benefit at work. You are also available . . . At the hospital [where I work] they ask that we have our own telephone so that they can call us when they need to be in contact. [. . .] We have established a system called "My watch" that is an internet system that assumes that nurses have a mobile phone. [. . .] And it sends out requests on shifts via PC and texting. [. . .] I don't use the system that much. It depends on whether there is a need for people for the different shifts, you know extra shifts you will have. We actually don't have that much in my department. [. . .]
>
> Interviewer: But that means that they use the system to get in touch with extra people that are not at work when others are sick and things like that?

Inga: Yeah. Then I can choose who I want to ask and I can set up the shifts that are empty, and then they get a text if they have a telephone or they can find it on the internet and then send a message back if they can take the shift or not.
Interviewer: So you don't call to their homes and ask?
Inga: No, we usually don't do that. But I have called nurses, for example, but it is always on their mobile phones because they have it with them.

Rather than a somewhat informal coordination of a sales force in the field, the mobile phone was being used for the systemic coordination of shift work. Like Mumford's and Thompson's discussions of the clock in the factory, the mobile phone had become a device of control (Mumford 1963; Thompson 1967; Zerubavel 1979). It facilitated the functioning of the hospital, since there was a need to have specialized people "on call." If one anesthesiology nurse, for example, had called in sick at the last minute, then the people staffing the shift could call other anesthesiology nurses who might be able to take the shift. The mobile phone added flexibility and responsiveness to the system, since the administrators could call the substitutes directly.

By 2009, this type of functionality had moved to less specialized jobs. A teen girl from the United States in 2009 noted:

I mean sometimes if you are applying for a job, they'll prefer to contact you by phone. And sometimes you can get an opportunity, like if you don't have anything to do and this person, one of the employees has to leave, [. . .] you can take over their shift.

While the type of work is not specified, we can guess that it is a service job commonly held by teens. The dynamics for the teen are the same as for the nurse, but instead of dealing with critically ill patients, the job probably involves standing behind the counter at a fast-food restaurant, selling clothes in a store at a mall, waiting tables, or stocking shelves in a grocery store.

In the two decades since the first group interviews, the mobile phone has moved through the workforce and facilitated staffing and work logistics at a variety of levels. It has changed the system of coordinating work and workers. Starting with relatively highly paid individuals, moving through specialists of different stripes, and eventually reaching relatively unskilled labor, the mobile phone has spawned a new structure of coordination (or perhaps control). The ability to microcoordinate has made the management of employees more flexible. Further, mobile phones are becoming terminals through which meetings are announced, agendas are distributed, participants are reminded, and, in some cases with the use of GPS, physically guided to a location. It has changed the interaction between employees and employers by making the former individually available to the latter.

Political Action and Resistance

A special case of coordination is seen in the use of mobile telephony to organize social protests. The mobile phone is being used in the logistics of collecting and deploying protesters. The logic of individual addressability is played out in the mobilization of people. Protests in the Philippines, Korea, Ukraine, Spain, at the World Trade Organization meetings, at the 2004 Republican National Convention in New York, against Mubarak in Egypt, and an ever-growing list of political demonstrations have been organized via the mobile phone (Castells 2009; Castells et al. 2007; Paragas 2003; Rafael 2003; Rheingold 2002).[9]

The general pattern of these actions is that, in a particularly tense political situation, people use the mobile phone to agree on a place to converge in order to protest. They then send a message describing the event and the time to the names listed in their phones, the people receiving the message in turn send it to their contacts, and so on. This creates an ever-expanding communication tree. For example, if each person sends the message to twenty unique others, within five iterations over 3 million people will have received the message.[10] This was seen in the "Arab Spring" of 2011 (Bhuiyan 2011). Castells (2009) also describes how, in the wake of the Madrid bombings, the Spanish electorate used text messages to organize their efforts to replace the conservative government. In this case, the call for support was dispersed via text messages. Another variation characterized the protests at the 2004 Republican Nation Convention: during an ongoing protest, people sent updates to one another on where they were needed to support ongoing actions, as well as places to avoid for fear of arrest.

The mobile phone is, however, a tactical and not a strategic tool. For the mobile phone to be a useful means of mobilization, there has to be a preexisting tense situation and also some broad sense of where and when the protest is to be carried out. Thus, the mobile phone is more of a fuse that helps to ignite a force that is already primed to go off. When there is a broadly recognized focus for the protest, the tree structure of message diffusion facilitates mobilization.[11]

At the same time that the mobile phone has been becoming a tool for the organization of social protest, authorities have started to develop measures to reduce the effectiveness of mobile communication in social protests.

Ang et al. (2009), for example, describe how the authorities in Nepal closed the mobile system during portions of the Maoist insurgency and how the Taliban in Afghanistan blow up base stations in order to disrupt the collection of information. The effort in Nepal was to hinder the logistics of the protest. Ironically, Ang et al. document that a major victim of the blockage was businesses and not the insurgents. A similar ban in Iran after the 2009 elections meant that people started to use the Bluetooth functionality of their devices to distribute information to people nearby. During protests in Iran, people uploaded footage showing the death of Neda Agha Soltan—a woman who was shot and killed—onto their mobile phones. This was then distributed via Bluetooth to other devices. In the next move of power politics, the authorities outlawed the use of Bluetooth. And so it goes. The protesters appropriate a technology, and then the authorities attempt to limit the functionality of the device if it runs counter to their interests. This escalation shows that protests and the response to them have moved into a new dimension. Information dissemination is no longer done using poorly mimeographed sheets handed out on street corners, but wirelessly via mobile devices (Castells 2009).[12]

Thus, as with the other forms of coordination discussed above, the mobile phone has changed the way that protests are mobilized. It is used to assemble protesters through their use of the device, and it is also used to direct the movements of protesters after they have been mobilized.

"Beeps," "Flashes," and Missed Calls as a Form of Microcoordination

As we have seen, in the developed world, mobile communication has altered the social ecology of mediated communication by changing the way we coordinate everyday life. In the global south, people have learned to use the mobile phone and its caller-ID functionality as a free signaling system that is used for coordination as well as expressive interaction (Donner 2007; Geirbo, Helmersen, and Engø-Monsen 2007). This is done by sending "missed calls"[13] as signals to one another. If, for example, a person at work need to signal his or her partner that they are done at work and needs to be picked up, they can send a missed call. That is, they dial the number, let the phone ring once or twice, and then hang up. Because of caller ID, the person receiving call knows who has called since he or she has coded the name into his or her phone's memory. The person receiving the missed call has a general idea that his or her partner, for example, needs to be picked up at a certain time; the missed call is simply a signal that he or she should start making their way toward the pick-up spot. This ad hoc signaling system is a simple and free form of microcoordination.

Missed calls have been utilized since the beginning of telephony. In landline telephone systems, where the caller was charged by the call or by the minute, this form of signaling was used in order to avoid charges. However, the fact that people can signal directly to an individual in real time has caused the practice to explode in developing countries. According to Geirbo et al. (2007), during peak hours missed calls can make up as much as 70 percent of the traffic volume moving through the mobile phone network.[14] In many cases, missed calls have the role of texting as a simple, mobile way of getting in touch and interacting with others. Missed calls are a new element on the scene that has replaced other more awkward forms of coordination and planning, and they have changed the social ecology of mediated communication in developing countries.

Although only a minimal amount of information is communicated, and while it does not allow for iteratively refining the agreement (the general meaning has to be agreed on beforehand), missed calls are extensively used. Missed calls can be used in a wide variety of situations and can have a wide variety of meanings. A missed call can often mean "come and pick me up," as noted above. However, it might also mean "please give me a call," or "lunch is ready" (Donner 2007). In the case of some trades, it can also be the signal of a potential job. Plumbers or carpenters might get missed calls with the expectation that if they want a job, they have to accept the cost of calling back.

Missed calls have been integrated into the flux of everyday interaction in surprising ways. Donner (unpublished) has reported on a sign warning people away from parking at a certain spot. The sign continues by asking people to send a missed call to a certain number if they have any questions. Presumably the person guarding the parking place is willing to pay for a return call if people have questions. In other instances, missed calls can be used as a sly (and cheap) way of telling a boy- or girlfriend that their paramour is thinking of them. In this case the interlocutors may not be able to describe the bounds of their feelings with just a single missed call. Indeed they may send hundreds of missed calls to a person throughout the day as if to repeatedly confirm the strength of their ardor. Geirbo et al. (2007) documented different individuals sending hundreds of missed calls to a single number. In one case a person sent over 1,000 missed calls to a single number. In qualitative analysis, they were able to confirm that this was a type of incessant valentine. In other cases, one user had generated 100 missed calls to 75 different numbers. This may have been an impromptu coordination of some commercial activity, such as taxi drivers who were asked to call a central number upon receiving a missed call.

The practice arose as a way to save the cost of calling, but it has since spread to all portions of society in some countries. It has become a generalized way of signaling one another, as with, for example, an affluent person summoning his or her driver to come pick him or her up.

Missed calls only function when both parties understand the meaning of the call. Nonetheless, sheer numbers show that they work. They fill a need. Missed calls are shorn of nearly all social courtesy.[15] Missed calls are a simple, direct way to get the job done. Missed calls have changed the way that many people manage and organize their social lives. They have replaced other more clumsy forms of interaction.

The different forms of mobile-based microcoordination (calling, texting, and sending missed calls) are more nimble than traditional time/place-based planning. This is not to say that time and place are not a part of the coordination picture, but that the mobile phone adds a flexibility and nuance to coordination that has not been a part of our social interaction before now. This can be good, but it can also add indeterminacy to our activities. Before mobile telephony, accepting the invitation to meet up at Cafe Cool at 8:30 with the group has meant that we potentially forgo other possibilities because of our loyalty to the group. However, having the ability to perpetually refine and negotiate agreements changes this. Instead of Cafe Cool at 8:30, we might suggest Restaurant Dud at 8:36.

The individual addressability of the mobile phone means that agreement planning is iterative. Microcoordination helps the small group of friends or partners decide where to meet and when. Microcoordination is used for the adjustment of prearranged meetings (i.e., those that have, as a point of departure, a fixed time and place but need to be adjusted because of exigencies). In addition, mobile communication allows iterative planning where the time, the place, the participants, and even the activity might be changed several times before the people actually meet up. All of this means that we have to negotiate our commitment to our social sphere. Rather than being a commitment, the gathering can be seen as optional or as something that will be "firmed up" as the time approaches. If there is a better (or more pressing) offer, we might renege on one agreement and substitute another. Where before our jilted meeting partners would be left to wonder as to what became of us, now we can send them a sheepish text with something that is perhaps more or less the truth. When we do finally meet up with the group, we might again be called away with yet another communiqué. In this type of mobile phone use, we are in contact with other group members and we are working out the nature of our social ties.

Just as with the regime shift seen during the horological revolution with clock time, microcoordination, in its various forms, is a change in the social ecology of mediated communication. As developed in chapter 3, the mechanical clock replaced alternative forms of timekeeping. The system of regular hours and minutes for marking the beginnings and endings of different activities replaced the use of idiosyncratic bells, position of the sun, and other time-marking instruments and systems. The mobile phone, in turn, is adjusting our reliance on the system of abstract hours and minutes. We are, in some but not all cases, coordinating our interactions with others through direct interaction and negotiation, and not in reference to the movement of the clock. Instead of setting up a meeting for coffee at 11:30, we might tell our friend that we will touch base later in the morning. When our respective responsibilities allow for it, we can agree to meet at the normal coffee shop as quickly as we can get there. This form of coordination does not directly rely on reference to clock time. It is a change in the social ecology of coordination that is, in many ways, suited to a society that bases itself on individualized transportation in a sprawling suburban setting. This is a theme to which I will return in the final chapter.

"We Ran Over a Girl and It Was Good to Have a Mobile Phone": The Mobile Phone as a Safety Link

In addition to facilitating microcoordination, mobile telephony has changed the social ecology of safety and security. In the case of both large and small incidents, the mobile phone allows us to contact others and to organize a response. In the 1995 group interviews, a respondent related the following, somewhat dramatic story.

Interviewer: Can you tell me about some of the situations where it is sensible, it is good to have a mobile phone?

William: If your job demands it, it is useful [. . .] in a crisis situation. That is the only thing. I don't see any need to have one as a private individual.

Anders: It is almost like, if you are way out in the wilderness and something happened.

Mats: We had a camping trailer one time and we ran over a girl and it was good to have a mobile phone. [. . .] It can be good to call for assistance.

The same urgency can be seen in the comments of a Norwegian mother in a 2003 interview. She told how her son had fallen off a dock onto some rocks. A passerby was able find the child's home phone number on his phone and was also able to call the local hospital. The mother said, "Without a mobile phone, I am completely desperate if I think about that. And skiing, he likes to ski a lot, it can be a life-saver."

This kind of use of the mobile phone can also be seen in the work of Cohen, Lemish, and Schejter (2007), who describe mobile phone use after the detonation of suicide bombs in Israel. Their analysis of mobile phone log data shows how people used the mobile phone to check on the well-being of their family and friends. They were able to trace the use of mobile phones in the immediate vicinity of the bombing, in a wider area, and in the whole country. They found that there was a dramatic peak in traffic near the incident as people frantically sought information on the status of family and friends. Cohen, Lemish, and Schejter (2007) also describe how the mobile phone allows people to alert the emergency services when there has been a bombing event. The mobile phone allows the authorities to be more responsive and helps them to organize the rescue operation. They are able to determine the location and the extent of the situation by talking to people on the spot. More recently, the mobile phone was used in organizing the rescue after the 2011 bombing in Oslo and the tragic events on Utøya, when teens called and texted the police and their parents. In cases such as these, access to the mobile phone was an important link to the authorities and also to our nearest friends and family. Analysis of telephone traffic following the Oslo bombing shows that in the first minutes after its detonation most people called their closest friends and family (Sundsøy et al. forthcoming). These analyses vividly illustrate how the mobile phone has become an important link to our most intimate sphere.

The mobile phone has become the tool of choice for immediately contacting our loved ones. We are able to get in touch with them directly; there is no need to call the police or hospitals when seeking information. The mobile phone has become the conduit through which we are able to assuage our fears, and in some unfortunate cases, through which we are confronted with the worst possible news (Dutton 2003; Katz and Rice 2002). It is an important link when people are faced with disasters such as earthquakes, typhoons, bombings and shootings.[16] People who have a mobile phone with them at all times can be warned about ongoing emergencies. In the wake of the shootings at Virginia Tech,[17] the tsunamis of 2004 and 2011,[18] and recurrent cyclones and flooding in Bangladesh,[19] SMS has been used to warn people in a given area of the danger they are facing. In Japan, for example, firefighting authorities have offered hotels software that would send text messages to guests in the case of a fire.[20] Personal addressability is at the core of these applications. It is changing the social ecology by allowing authorities to transmit important information in critical situations when other information channels are not available.

Mobile telephony is changing the parameters of how people are informed in these dramatic situations. Luckily, not all emergencies are quite so dramatic. Haddon (2004) suggests that our interpretations of emergencies can vary. An emergency can be a physical injury, an earthquake, a car breaking down, or perhaps a less serious disruption of daily life. Our use of the mobile phone in these situations illustrates how it is changing the social ecology of mediated communication.

In both large and small ways the mobile phone has changed how we deal with actual or impending emergencies. It has allowed people who are lost or injured in the outdoors to contact rescuers.[21] It has helped people who have met with catastrophes, such as floods, acute medical problems, or car accidents in remote places, to seek help. These examples illustrate the range of situations in which we use the mobile phone. The ubiquitous access afforded by the device is a real advantage when compared to technologies that are more place bounded. Landline telephony suffers in this comparison, as do traditional desktop PCs. The accessibility provided by the mobile phone has changed the way we organize responses to disasters. It has provided a new logic to the way we approach safety and security.

"She Was Calling Me Like Every Day": Parent-Child Relations in the Age of Mobile Telephony

On the whole, the mobile phone enhances the cohesion of the closest sphere of family and friends. It gives us better, more direct contact with family members, and thus it give us a sense of security and provides a sense of cohesion. Research suggests that the mobile phone is enhancing social cohesion in small groups and, in particular, the family (Ling 2008b). Kim et al. (2006) in Korea, Matsuda (2005, 127) in Japan, and Smoreda and Thomas (2001) in Europe point to the importance of the mobile phone in family interaction as well as friendship networks. Further, Donner (2005) in Africa and de Gourney (2002) in France describe the same general trend. Matsuda's (2005, 133) "full time intimate sphere" and Habuchi's (2005, 127) tele-cocoon give us a sense of how the mobile phone contributes to the social integration in the closest sphere. Further, Castells et al. (2007), Miyata (2006), Igarashi et al. (2005), Ishii (2006), Stald (2007), and Reid and Reid (2004) all describe different dimensions of cohesion being fostered by the mobile phone. Licoppe's (2004) notion of connected presence also plays on the same general theme. The mobile phone clearly has changed the way that parents and children interact with each other. It provides a (potentially) open conduit between them that can be used to facilitate practical matters, or to impose into the others' sphere at unwanted moments.

The Mobile Phone and Parental Expectations

Using a mobile phone changes the mediated interaction between children and their parents. It affects coordination, the way that security is practiced, and the child's sense of independence. When seen from the perspective of the parents, there are issues of safety and independence that affect their decision to give their child a mobile phone. From the perspective of the child, the mobile phone is also changing the landscape. In this case, the focus is on autonomy and self-sufficiency.

Parents typically buy their children phones so that they will, for example, not be marooned without transport. They also buy them to simply facilitate complex family coordination. In some cases, children who suffer from sickness will get a phone so that they can call their parents if an acute need arises. In other cases, children of divorced parents will get a phone so as to facilitate access to the nonresident parent (Hjorthol et al. 2007).

As children enter middle school and a much wider social world, there is often pressure to outfit the child with a mobile phone. In other cases, parents buy phones for their children when they go on trips. Material from the United States in 2009 points in this direction:

Sally: I was 12 and I got it because I was going to New York by myself, and it was the first time I was flying solo. But, of course, when I got to New York, I dropped it in a pond in Central Park, so [group laughs] it still worked after that, so that's cool.

Sally continues that her increased accessibility also had its difficult moments:

Sally: I guess it is two years ago now, I went on a lot of trips like with school and stuff and they would . . . we went to Quebec and it was a dollar a minute [to call], so I didn't bring my phone, and it drove my mom nuts that she couldn't call me. But then, when we were in Chicago, she was calling me like every day.

As reflected in Sally's comments, once the child has a phone, they are often expected to grant parents access regardless of the situation in which he or she may be. Not being available is a cause for concern. Sally tells of another incident where her mother became frantic when she was not able to contact her:

Sally: Well, one time, my mom usually calls me when I get home from school, and I didn't have my phone that day and the house phone was being weird, and she like freaked out and called one of my friends and was trying to figure out where I was.

Joan: That happens a lot with my mom.

Interviewer: Yeah? It does?

Joan: Yeah, she's like, "the reason why I got you a cell phone is that you can call me when you get home or when you get to this place," and if I don't call she like flips.

"Why do you have a cell phone if you can't call me?" Like, "I got caught up." She'll be like, "no, no excuses." So, having the cell phone is a pro and con.

The rules governing a child's access to a mobile phone becomes an arena where parents and children need to develop an understanding. There is, however, a sense among many parents that so long as it is a safety measure, the mobile phone is a positive thing. As we saw in chapter 6, when mobile phones were first becoming available, a teen was seen as being cheeky if he or she had a mobile phone. However, this quickly changed. Mothers in a 2003 focus group in Oslo were grateful for the contact that the mobile phone gave them:

Maren: I could not have managed without a mobile phone.
Andrea: It is a completely different way of dealing with children today than before. My daughter got a mobile phone when she was 16. Before that, I never knew where she was. After the mobile phone it is completely different. I have not been afraid one day since she got it.
Julie: My son will be 15, and he is on the point of changing, that he wants to go out and prove himself a little, you know. So he calls from his mobile and he says that he is at John's house, but you don't know if he is at John's house.
Maren: Then I call John and ask to talk with him. I always do that.
Julie: Yeah, I also do that . . .
Andrea: Yeah, but they might be out and about and not catch the tram and come home too late, and then you are afraid. If you have a mobile phone, then you have the chance to call them. Before, I rushed around, I remember when I was 15 there were not mobile phones, and we looked for phone booths.

When comparing the situation of their youth to that of their children, these mothers appreciate the accessibility that the device gives them; at the same time, they are the last generation that grew up without mobile phones, and thus they have a special perspective on how it has changed parenting.

"She Cannot Tell the Difference between What Is Important and What Isn't": Cutting the Cord with Children

There is a paradox associated with children's use of the mobile phone in the eyes of many parents. On the one hand, it gives parents direct access to their children. There is no need to wonder when they are coming home for dinner or where in the neighborhood they might be. Also, the mobile phone allows for remote parenting for parents who work when their children have come home from school (Rakow and Navarrow 1993). It can, however, also provide too much access (Chen 2007). There is sometimes the sense that children are not learning how to make their own decisions. This is seen in the discussion of adult women in a 2003 group interview:

Anne: [. . .] a typical mobile phone [call was that] our daughter called and asked if she could have some popcorn.

Interviewer: That was a typical conversation?

Anne: "Can I have a stick of gum?"

Interviewer: What would she have done if she didn't have access to a mobile phone?

Anne: She would have taken it anyway . . . Either taken it or not taken it.

Grete: But then she would have been more independent, maybe, and made the decision herself. Isn't it a little dumb? "Can I have a piece of gum?" Honestly, I give up.

Ida: It is a little sweet also.

Anne: Yeah, but she doesn't call for more important things. She cannot tell the difference between what is important and what isn't.

Ida: Maybe she wants contact.

Anne: It isn't that, you know. She isn't interested in talking. "Can I have some gum? Yes or no?"

Grete: "Great! Goodbye then."

Ida: She wants to have contact with you.

Anne: Actually it is more than that. She is home alone because her father had to do something this evening. So it is more than just popcorn, you know.

Ida: Then she is not so alone at home. That is why the mobile phone is good. A little company.

Anne: Yeah, but I think that children have become less independent.

Tanja: I think so also.

Anne: They have to learn to make decisions. Maybe it's a wrong decision, but they learn to trust themselves.

This dialogue shows that the women were not able completely to make out the meaning of the phone call asking for the popcorn or gum. On the one hand, it was a typical message associated with the ongoing maintenance of the home. The child may have thought that the popcorn or the gum was being saved for some special occasion. As noted by several of the people in the group, the call also may be a search for contact and reassurance while the father is away on business. It may well be that the child is not comfortable being home alone or assuming so much responsibility. The mobile phone is a simple way to clear permission, or it may be a way to reassure herself that her mother is actually available. This latter interpretation, in turn, brings up the issue of how independent a child should be. The fact that parents—and following Chesley (2005) and Rakow and Navarrow (1993) it is particularly the mothers—are always only a phone call away may suggest to parents that their children are not being independent enough.

Parents of teens often express the need to let their children experience the world. In doing this, they experience a tension between guarding their children and giving them the space with which to develop into an adult. The mobile phone gives them a tool in this side of life.

A mother from Oslo who participated in a 2003 group interview reported that she told her daughter, "If something happens, call home and we will come immediately!" The mother went on to say that her daughter also needed to experience the world and to learn how to deal with different situations. Other mothers also discussed this paradox.

Interviewer: For those of you who have daughters and sons also, when they start to be teens and start to think about being on their own: suddenly they disappear on a vacation without you and things like that. Do you think it happens earlier now than before since they have such easy access to communication?

Vilde: It is not earlier, but maybe it is easier for parents to let them loose because you have the possibility for communication. It doesn't have to be earlier.

Frida: I think that we are more afraid now than before. If you haven't heard something for half a day then we are afraid. It is not so dangerous. When I was 18, I was on InterRail for a month and called home twice.

Tuva: I sent a postcard.

Oda: Yeah, I sent a postcard . . . The postcard from Spain arrived after I came home.

Maja: My daughter calls home every day.

Frida: But that means that we don't trust our children as much that they can make it [on their own]. We as parents think about our own security, parents always do that. For us it is very secure, but it is not given that it is best for the children that we always call them.

Vilde: But it must be an advantage for the children [to have a mobile phone] also. They can get into situations where they need help.

Karoline: But we did that also. It has always happened that we got into situations where we needed help. But there is not a higher death rate now.

Marta: I grew up with parents who didn't have a [landline] telephone. I don't know if it was based on principle or what but they didn't get a phone before I had moved out. It was really difficult sometimes. One time I got into a mess, when I was 13 to 15. It was not nice at all, but I don't think that it made me into a bad person. It would have been good to have a phone.

As with the previous group of mothers, the participants raise the question of their children's independence. On the one hand, the women are thankful for the contact, but they also have concerns regarding their children's independence. Parents and children can more easily maintain contact, and it is easier to react should a situation arise. However, as is obvious from the comments of the women interviewed here, they are also worried that their children will not gain the ballast that they need to be out in the world. If they can always call for help, they will lose their ability to act independently. At one level, these women are not really commenting on the situation of their children, but rather how the mobile phone changes the way we think of security and safety. In their youth, they had gone off to far shores with little chance of getting in touch with their parents; their children are living in a different world. Mastery of independence may be in the process of being replaced with mastery of mediation. Indeed, Chen (2007) finds that this tendency can extend into the university years.

"I Don't Talk to Her Too Much": Children Calling Parents

Parents use the device to be in contact with their children. For their part, some children use the phone to maintain the emotional bond with their parents. Terry, a teen girl from the United States in 2009, said, "Well, I usually text my mom like 'so, how are you doing?" stuff like that." However, it is perhaps more often the case that teens see the mobile phone as a key to their growing independence. As noted earlier by Amber, using the mobile phone makes her feel "older and independent." This was an important consideration for her. With this growing independence, parents can be placed in a more peripheral position. This is seen in the comments of Trevor, a teen interviewed in the United States in 2009:

> We have like the "fave fives" and we used to like have my Mom on it. But it seems like I don't talk to her too much so I took her off, but the only reason why I usually call my Mom is like if I want to ask her something or if I want to go somewhere. Or, like, what time I need her to come and pick me up. But, she interacts with us like just to see where we're at, to be in our business.

Trevor's begrudging placement of his mother as a useful resource, but not a central part of his social sphere, is perhaps indicative of his growing emancipation. His mother is still a central part of his life, and she is a facilitator, but his comments show that his social horizon is expanding. Trevor's mother is not worth having as one of the more central persons on his phone. His comments illustrate that during adolescence, friendships perhaps more than family are the central relationships in a person's life. Up to adolescence, the family has been fundamental; after Trevor enters into adulthood and eventually marries, the family of procreation will command a vital position. During adolescence and young adulthood, it is friends who have this position (Rubin 1985). This is seen in his organization of his contacts on his mobile phone.

Thus, the accessibility represented by the mobile phone cuts both ways. On the one hand, teens have access to friends, but parents also have access their children. There can be tension in this interaction. Trevor's mother is in all likelihood a major source of logistical support (driving, cooking meals, washing clothes, etc.), but he also thinks that she is "in his business" too much. In all likelihood, Trevor's parents are also unsure as to how intensely they should impose themselves into his life. As seen in the last section, parents are often unsure as to whether their call is a welcome point of contact or if it is an unpardonable intrusion—if it is needed support or if it robs their child of independence. Thus, the mobile phone can become the locus of an uncomfortable interaction between parents and teens. In one turn, the contact is necessary and indeed welcome, but in the next, it is painfully close. In the one instance, both parents and children can enjoy a quick chat or feel good about being driven to one or another activity. In the next cycle, the leash feels too short. Regardless of how this is arranged in different families, the mobile phone has changed the social ecology of mediated interaction.

The mobile phone has established itself as a link between family members. It functions to tie the family together and to assist them with coordination and in the expressive work of family interaction. It has changed the way teens experience adolescence and the way they interact with their parents. This is not to say that other forms of mediation are not a part of the picture. Social networking sites, email, landline telephony, and other forms of mediated communication are still in the picture. However, when focusing on point-to-point interaction, the mobile phone allows teens a form of independent action. It allows them to coordinate their activities better and it gives them more direct access to their circle of friends. In addition, the mobile phone gives them phatic and functional contact with their parents. It allows them to summon parents when needed and it can be used to manage the awkward requests of the parents. It is changing the way parenting is done, and it is changing the conditions of its practice. Because of the mobile phone, the radius of action for both the parent and the child is somewhat larger and more robust. The link can be abused by overly nosey parents and/or overly secretive children, but both know that, if needed, the other is only a call away.

Texting as a Low-Overhead Form of Interaction

Texting, and to some degree mobile use of social networking sites, is another way in which the mobile phone has changed the social ecology of mediated communication. The mobile phone gives us individual addressability; we can interlace these actions into our daily routines influences the way we think about safety, coordination, family life, or adolescence. In each of these, there is a large component of voice interaction. However, it is worth considering how texting has changed communication. It allows us to fire off a simple message that the receiver can deal with as time allows.

In chapter 5 we saw how texting has grown from nothing to being one of the dominant forms of mediated interaction. A text is easy to send. It is often shorn of greetings and courtesies. We dispense with much of the etiquette associated with extensive salutations and elaborate closings. Indeed, only rarely do we use them in texting (Ling 2005; Ling and Baron 2007). Texting can be used to communicate a short point without the need for elaborate social overhead. It can be like a simple digital note, composed quickly and sent into the ether. Because of these characteristics, texting has found a secure niche in the communication ecosystem.

Texting is often more direct—some would say abrupt—than, for example, voice interaction, since in voice communication and in other forms of written interaction we have developed elaborate forms of greetings and closings. There is, however, a sense that texts have become the "post it" notes of mediated social interaction. Teens in the United States note that it is a quick, convenient way to interact:

Sally: [. . .] if it is going to take me less than a sentence to say, if I just have on little quick thing, I will just text them . . . Texting is just like "can you do this?" "OK."

We can send quick messages to one another, which can be used strategically in order to ask quick questions. According to another respondent in the same focus group, "texting is really handy if you have to ask somebody a question, but you know that if you call them it is going to be at least an hour-long conversation."

Texting, as well as mobile voice interaction, often takes place among a relatively small group of individuals. Analysis of traffic data from Norway has shown that half of all text messages go to five or fewer other people (Ling, Bertel, and Sundsøy 2012). For teen girls it is about six, and for mature adults it is about three.[22] In the case of teens, however, these few people are the objects of intense interaction. In a sense, there is ongoing interaction throughout the day.[23] But regardless of age, texting is used to maintain contact with the core group of family and friends. Texts might focus on deciding who will carry out different tasks (shopping, picking up and delivering children, etc.), and there can also be phatic interaction, to occasionally check in (or check up) on a partner, a friend, or a child (Lasén 2011). Regardless of the content, there is a steady stream of texts flowing back and forth.[24] Some of the texts may be dramatic, but largely they are mundane. It is in these bits and pieces of interaction that we can send our partner a status report, get a greeting from a friend, or receive a request from our child to be picked up. Through these humdrum interactions, however, we develop and maintain social bonds and we work out the activities of daily life. These small bits of interaction have given us a new forum in which we can be social.

"It Is Easier to Send a Text": Social Interlacing of Mundane Activities

One of the ways that texting has changed the communications ecology is that it can be done while at least ostensibly participating in other activities. As Leopoldina Fortunati (pers. comm.) so nicely puts it, texting fits into the folds of our lives. It allows us to use otherwise unused moments to get something taken care of, or to give someone what Ito and Okabe (2005) call a little tap on the shoulder. An interviewee from 2003 said, "You can sit on the bus and think that it is a long time since I have had contact with this or that person, so you can just send a text when you are sitting there anyway." Texting allows those who are socially dexterous to attend to two publics: the copresent and the mediated.

The interlacing of communication can be seen in many situations. Indeed, there is an "under the radar" ubiquity to the use of texting. In one of the US focus groups to investigate the use of mobile communication, teen girls actively texted and maintained their social interactions *during* the focus group. As the interview developed, it was easy to see that many of the girls had their mobile phones in their laps turned to silent mode. From time to time they checked for messages when other girls were being called on or were talking. These teens were able to interlace the focus group discussion and the mediated interactions with grace. They responded to questions in the focus group and contributed to the discussion, all while they were contributing to the trillions of text messages being sent around the world. When they were not actively engaged in the group interview discussion, they would check their phones for new messages and perhaps type out a quick reply. When we saw this, we asked how many of the participants had received texts:

Interviewer: How many of you have gotten a text message since you've been here?
[Laughter]
Sandra: I got a phone call, but then I hung up.
Interviewer: OK, so how many have you gotten since you've been here?
Teresa: I got like four in a row just a few minutes ago.

Nearly all of the teens had. Most said that they said that they had received 5 to 10 texts in the 45 minutes they had spent in the focus group. When asked about the content of the messages, they told us that the texts were updates about their friends' activities and about what the girls would be doing after the session. All the while, they maintained their participation in the ebb and flow of the copresent interaction.

The group interview had gone from being the study of something happening "out there" to being a type of participant observation. The girls were interlacing their copresent activity with the planning of the next stage in their social lives. They were keenly involved in the flux of their friends' situations; all the while, they were carrying their part in the group interview. They were able to glance at the phone, or to tap out some words in stolen moments, but then would quickly reengage in the flow of the discussion. In one turn, they were deciding where to meet their friends in an hour's time; in the next, they were talking about the rules of mobile phone use at school. They were picking up a bit of gossip from a friend, and then they were telling the researchers about how they never turn off their phones. The mediated and the copresent were interwoven into a piece. They could set aside the texting, and their attention was drawn back to the copresent situation. This type of communicative multitasking illustrates how the mobile phone has changed both the rules for accessibility and the social ecology of communication. An "event," such as the focus group or class or in some cases a family dinner, is not a bounded affair. The mobile phone has restructured events so as to make us perpetually available to one another (Katz and Aakhus 2002).

Texting took off in the late 1990s and soon gained a foothold in daily life. By 2003, texting was being used to arrange social interaction and to maintain contact with friends. Young adult women discussed how they use the device to send simple greetings and updates to one another with the mobile phone. This comes out in a focus group among adults in Norway.

Interviewer: You all use [SMS] as I understand it. You send a lot to friends and husbands and boyfriends and such. What are these messages about usually?

Elisabeth: "Do you want to go to the movies tonight" or "Is there anything happening?"

Kathrine: . . . or something funny happens right then, then I have to send a text. If there is something fun that happens and you send it along then that is very nice . . .

Interviewer: [. . .] Have you had more contact with [friends] after you got your mobile phone or is it the same?

Elisabeth: There is more contact. "How are you doing" and "What did you do over the weekend?"

These women describe a more-or-less continuous interaction with friends and family. Interaction is not only reserved for those times that they are copresent; rather, it continues interlaced into other activities when they are not physically together. They are able to share mundane interactions, about things that have more or less import. The interactions are not just about relaying a bit of information; they are about keeping the communication channel open. As Licoppe (2004) has suggested, this type of contact lets us maintain a type of ongoing interaction, what he calls "connected presence," with significant members of our intimate sphere. Where before we might have waited for a more extended time to talk to one another about the large and small issues of daily life, texting and talking on the mobile phone have become an ongoing form of interaction. These small updates and communications slip into the flux of daily life.

This type of interaction is also being facilitated by access to the mobile internet and social networking sites. We are able to take a quick glance at Facebook for updates in our social network and messages from friends in different groups. In many cases these are not actionable communications, but rather simply interesting bits of flotsam and jetsam that orient us vis-à-vis the activities of our social crowd.

Texting and social network updates do not demand the simultaneous attention of the interlocutors. We can simply leave a message, or, if our texting partner is available, we might exchange a sequence of texts in a type of text-based "conversation" to coordinate possible interaction or share a few friendly comments (Helles 2009). If a friend is not able to reply to a text, at least we have contributed our side of the conversation by sending along the information. Calling can disturb the other person and often involves more elaborate greeting and closing rituals. Email is often limited to important work-related things, especially among teens. A PC with access to Facebook is not always handy. Thus, texting via the mobile phone is the least common denominator for communication.

Texting helps us to deal with mundane interactions. Such interactions, by themselves, might not be of sufficient moment so as to demand synchronous discussion along with all the social encumbrances of opening courtesies and small talk that lead up to the actual motivation for establishing contact (Ling 2005). Rather, a text can be received, read, and responded to, quite literally, under the table. A text does not need to disrupt the flow of other copresent activities. This more casual style also means that there is a lower threshold for sending texts:[25]

Sunneva: It is easier to send a text that says "Hi, How are you doing" or "Do you want to go to the movies" or some other embarrassing thing. It is more innocent. To send a text, anyone can do that. It is easier to interrupt [someone] with a message. You can't always do it either. If you call, it is in a way more personal, more direct. If someone is asked to go to the movies then with someone they don't want to go to the movies with, it is a lot better if you write that. "No, I can't today because I'm doing something else." If you call it is more like aaah, you know. (Young woman in the 2003 Norwegian group interviews)

Texting allows a person to participate in one setting while physically being in another. Further, it is an immediate and easy way to deal with the small stuff of everyday life. It can be used to deal with issues where there is no need for a long, complex discussion. Messages via text or mobile social networks can be slipped into the flux of other activities with little pretense. Our willingness to interlace them into other copresent events indicates that they are a new element in the social ecology.

"I Often Send a Text Message When I Should Have Called, but I Can't Bring Myself to Do It": Managing Interaction in the Texting Era

Since it is a less imposing form of interaction, texting allows us access to others that would not have otherwise been possible. It allows us to better control the flow of interaction. For teens who are in the process of forming relationships, texting allows them "time to think" as they compose their flirtatious interactions. In other cases, it helps them to keep the conversation flowing and, when they run out of topics, to end the interaction:

Rick: Some people are bad at talking on the phone, like some people just really want silence and don't really want to talk. Like when I text I can just say what I want to say and I don't have to constantly be talking. If I have nothing else to say then I will just stop texting you. Or if we just run out of stuff, we just stop texting. Whereas on the phone you try to just keep on [talking] . . . I mean on the phone, you always try to make the conversation go longer—if you are talking to somebody, not if you are calling to ask for something. So, it depends on who I am talking to. (US teen interviewed in 2009)

As Rick's comments suggest, voice telephony is more unwieldy than texting. Voice interaction requires management of pauses as the interlocutors think on their feet. Rick obviously felt uncomfortable with this aspect of voice interaction. Texting allows interlocutors to think between utterances. It also gives Rick an easy way out of the interaction. He notes, "if we just run out of stuff, we just stop texting." There is not necessarily a need to develop excuses and ruses in order to back out of a social interaction. A person who is rounded in their texting courtesies might send their texting partner a thin excuse when they want to turn to other activities. They might text that they have to go to class, or eat, or whatever activity provides them with an exit. According to Rick, however, even this small courtesy can be dispensed with.

Texting can also be used strategically to decline invitations. This does not mean that the messages are less urgently felt; it only suggests that there is a lower level of confrontation associated with the eventual rebuff. The comments of Ayesha, a teen interviewed in the United States in 2009, also suggest that there is a strategic dimension to texting.

I text my sister rather than talk to her because I'll be afraid of the answer, so I would just like . . . I don't know, I'd rather text my mom if I have something to ask her. Like "so, can I do this?" when I know she'd probably say no if I called. She's just like "sure, whatever" just to get off of the phone, you know? And I don't like texting my mom, because she'll be at work and she don't reply as fast I need her to . . . She be doing other stuff and it'll be taking like [a long time] . . . I ask her something and she'll text me two hours later like "k."

In this situation, texting is used either to avoid a direct confrontation or to slip past a potentially awkward interaction.[26] People reported using texting to avoid boring but necessary conversations:

Brie: [Texting] is good if it is people that you have not seen for a long time and you don't want to call, you can just send a text and ask how it is going and so you don't have to have direct contact.
Olivia: You also save time.
Talia: Yeah, you avoid sitting there for twenty minutes you know. (Adult women in the 2003 Norwegian group interviews)

As these women note, texting allows us better control of the conversation. This can, however, be seen as a cheap way of brushing off social responsibilities.

Emma: It is very easy to say via a text message that "I am thinking of you," but it usually means nothing.
Ava: It means that you have thought of them, though.
Abigail: That you have a bad conscience, for example.
Chloe: It is a question of what type of relationship you have with them, though.
Ava: I often send a text message when I should have called, but I can't bring myself to [do] it.

These comments show that texting can have an ambiguous character. Clearly, when the texts are associated with coordination, there is an instrumental logic to their interpretation. However, when there is the sense that they are being employed in lieu of other social interaction, they may have a more problematic character.

"She's a Little Slow on the Uptake": Parents, Teens, and Texting

Perhaps more than anything else, texting is a teen form of interaction. Although texting does take place between parents and children, it is not as well entrenched among parents as it is among teens. Analysis shows that the vast majority of texts are between same-aged teens (Ling, Bertel, and

Sundsøy 2012; Ling and Stald 2010). In some cases, only certain members of the family can or are willing to text.

Sarah: Um, I text my Mom sometimes, but I'd rather talk to her on the phone because . . . I don't know, she's not that great at texting and she's a little slow on the uptake. (Teen girl in the United States interviewed in 2009)

There is not just the sense that parents are only marginally able to text; in addition, when they do text they can be seen as less in control of the situation. Some teens have a sense of domination when texting. Further, teens are aware that not all people are able to text. Elderly people are often those who receive the fewest text messages.

Rachel: The only person I call in my family is my grandma because she texts really slow . . . Well, sometimes my mom texts me. She's like, when are you coming home? And I'll say "when I want to." She'll be like "okay, well, be home by eleven." And I'll be like "okay." And then my sister will be like "can I wear your clothes?" and I'll be like "no" and she'll be like "okay." And then my grandma, I'll be like "hey, can you give me a ride?" and she'll be like [old lady voice] "Rachelley???" She has to call me. (Teen girl in the United States interviewed in 2009)

Rachel's comments indicate that texting gives her a sense of dominance in the management of the interaction with her mother and sister. It is as though she feels emboldened in the use of text. It seems that when texting, she has control over the relationship, and her mother is the interloper. Rachel presents her mother as timidly asking when she will be home, to which Rachel gives a brisk reply. Rachel's sister receives an even more direct reply to her request. Rachel's argumentative comments describe a tension associated with the mobile phone. The mother is requesting information and, at least as portrayed here, Rachel is determining the conditions of compliance. The filter of the mobile phone gives Rachel a certain command. Previous generations would have had to call directly or would have had to have a specific agreement about when they would be home. Here there are negotiations, and they are carried out through a channel that perhaps favors the position of the teen.

As Rachel's comments suggest, voice interaction is more appropriate with the grandmother, but texting can be used with her mother. In general, we have a sense that texting is not a suitable form of interaction with elderly persons. There is indeed a sense that calling is more appropriate when interacting with elderly persons (Ling 2008a). Because the elderly do not generally send text messages, they contribute to the idea that voice is the preferred form of interaction. There is a question as to whether this lack of intergenerational texting, particularly with the elderly, is a result of poor usability, some type of exclusion, or perhaps both.

Texting is often shorn of elaborate courtesies, and it is brief. It can be used for quick messages, but it can also be used with grace and elegance. Its central function seems to be in the communication of quick "post it"-like interactions. Texting and other forms of mobile text-based communication exist in a part of the communication repertoire that prizes brevity and can happen in the background. It allows us to dispatch messages in stray moments that the receiver can deal with as the chance arises. Texting is an interpersonal form of interaction where we address our messages to a specific person. Texting has changed the social ecology of mediated communication by allowing for direct interactions that can be produced, read, and responded to on the fly or interlaced into other situations. Texting has given us a low-overhead form of communication.

Information Access and Mobile Telephony in Commerce

Up to this point, I have looked into how the mobile phone has changed the social ecology of the private sphere. Comments of businesspeople clearly show that mobile telephony has also had an effect on the way that business is done. It is clear that the internet has had dramatic effects in this area (Castells 2009). Traditional computing and the development of the internet have changed the dimensions of how commerce takes place. There are also cases in which mobile telephony has facilitated logistics and the flow of information in commerce. These can be situations where portions of the production process take place at distributed sites where there is not reliable access to a PC-based infrastructure. The Norwegian salespeople cited above clearly saw the benefits of being personally available at all times. They understood how the mobile phone had made their work more effective. The salesperson Oskar said that the mobile phone allowed him to cover a larger area and "always [be] in contact with the main office, and at the same time, you are in contact with the international suppliers." Another salesperson, Tobias, said that after he got a mobile phone he could do a more rational job of establishing his routes. The mobile phone has given traveling salespeople access to pricing information or changes in the location or timing of a meeting. The adoption of smartphones, the mobile internet, and location-based services is adding a new dimension to this discussion (Stuko and de Souza e Silva 2011).

Another function of mobile telephony is that it gives users access to time-sensitive information. In some cases, this has changed the way that commerce takes place. In its time, the landline telephone provided the same type of function. Fischer (1992) discusses how farmers in the United

States used the telephone in order to follow crop pricing and the advisability of holding or selling their crops at a given time. The spread of the mobile phone meant that this functionality was no longer located in a fixed location, and in the process, it has changed the functioning of both commercial organizations and local markets. This type of impact has now moved into the developing world.

One of the most exhaustive studies of this impact was carried out by Robert Jensen (2007), an economist who studies the interaction between local markets and access to price information. He was interested in how mobile communication affected the market for freshly caught fish in Kerala, India. To study this, he gathered information on the price of sardines (a staple food in Kerala's communities) at approximately the same time every day for five years, between 1997 and 2001, in fifteen different markets. He followed the pricing in three general areas. Each area included five "beach markets" where the fishing boats sold their catch. Each market was approximately 15 kilometers from the next. As fortune would have it, this period saw the introduction of mobile telephony to the area. Thus we are able to see the effect of better information flow.

Before the introduction of the mobile phone, the price of fish could vary dramatically from day to day. During this period, the fishermen usually fished near their own home harbor and returned to the same harbor with their catch. Since almost all the boats did the same, the access to fish varied for all the sellers/purchasers in that harbor. If the fishing was poor, then the price of those few fish that were caught was driven up. On the other hand, if the boats were lucky, then there were many fish to be had and the price was low. In some cases, the price was so low that there was no market for the fish from late-coming boats. In this case they simply dumped their catch. That is, they got no payment. While the price at one beach market might be high, the boats attached to the neighboring market might have been luckier with their fishing, and correspondingly the price per kilo for sardines in that market would have been low.

Looking at this in economic terms, the broader market across all fifteen harbors was not functioning optimally. It was metaphorically—and perhaps literally—either feast or famine for both the fishermen and the households who bought the fish. One day there were too many fish, and the excess was simply dumped. On another day, there were almost no fish to be had. Those that came in commanded a high price.

According to Jensen's (2007) analysis, the mobile phone changed all this. The fact that the boats could call into their home port and compare the price of fish to that in other nearby ports meant that the fishermen could choose the location where they would receive the best offer. They could go to a market further up or down the coast depending on the price information they gathered (or the agreements they were able to forge) while still out at sea. Jensen's material shows that between 30 and 40 percent of all boats delivered fish to beach markets outside their local market on any given day. At the same time, the purchasers on land also received benefits from the new system. Rather than being subject to the whims of the fishing banks immediately near their homes, they were able to rely on the fact that boats from other nearby areas could come to meet their needs.

The material in figure 7.1 shows the effect of the mobile telephone on the price of fish. The figure is divided into three panels, where each panel shows the price (in rupees) per kilo of fish over the period of the data collection. Region I was the first to receive telephone coverage in approximately week 22. The mobile phone coverage extended about 25 kilometers out to sea. Thus, it covered the major fishing areas. Region II received coverage in week 98 and Region III received it in about week 198.

The data show the effect of the information channel on the pricing of sardines. Jensen writes, "the addition of mobile phones was associated with a large and dramatic reduction in price dispersion and waste" (Jensen 2007, 900). The most striking aspect is that the variation in pricing is almost eliminated. The data show that pricing was a chaotic mess before the adoption of mobile communication. Afterward, it approximates linearly. Second, waste is eliminated. Since the fisherman always had an alternative market within range, there was never the need to dump their catch. Third, the stable pricing helped both the fishers and the buyers in that there was secure knowledge as to the expected price. This facilitated budgeting for both the buyer and the seller. There was a slight increase in income for the fishermen, largely based on the elimination of waste. Indeed, the payback time for a mobile phone was only about three months.

Thus, we have an example of technology introduction that seems to result in a better situation for all parties. The pricing of a staple food was evened out because of the introduction of an information channel. According to Jensen:

> We find that the addition of mobile phones reduced price dispersion and waste and increased fishermen's profits and consumer welfare. These results demonstrate the importance of information for the functioning of markets and the value of well-functioning markets; information makes markets work, and markets improve welfare. And it is again worth emphasizing that the results represent persistent rather than one-time gains since market functioning should be permanently enhanced by the availability of mobile phones. As mentioned earlier, information technologies

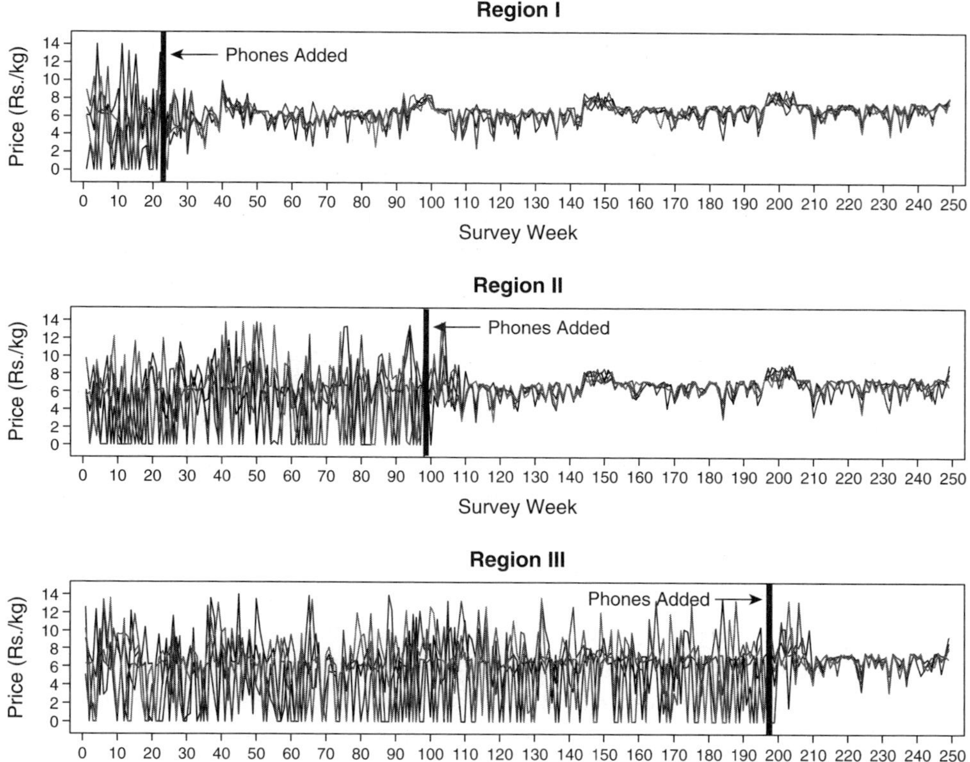

Figure 7.1
Sales price at beach markets (in 2001 rupees) for sardines in southern Kerala, India (source: Jensen 2007).

are often considered a low priority for developing countries relative to needs in areas such as health and education. However, not only can such technologies increase earnings, but those increased earnings (or increased purchasing power, due to reduced consumer prices), in turn, can be expected to lead to improvements in health and education. In addition, because mobile phones in Kerala are a private sector initiative rather than a development project, other than through perhaps raising interest rates for capital, they do not crowd out investments in other projects. Also unlike most development projects, the service is self-sustaining; mobile phone companies provide service because it is profitable to do so, and fishermen are willing to pay for mobile phones because of the increased profits they receive. (Jensen 2007, 919)

Jensen's work gives us a description of how the mobile phone can influence the workings of a market. The communication channel provides benefits to both the fishermen and the consumers. The accessibility and communication provided by the mobile phone have changed the social ecology of the fish market for these communities.

There are limitations to the situation. Jensen notes that the perishability of fish is a particularly important aspect of this market. The fact that refrigeration was not available meant that the fish necessarily needed to be consumed within a few hours of being caught. This aspect led to the system of dumping the fish before the introduction of the mobile phone, and it partially explains the success of mobile communication. Thus, we might expect that similar success stories could be observed in the market for other perishable commodities. Jensen suggests the market for milk, eggs, fruit, vegetables, and spot-labor markets might also benefit from more efficient information channels (ibid., 920). The fishing example also is somewhat special, since there are a large number of agents on both sides of the market. This hinders collusion in the market, and it means that the market is not distorted by powerful actors.

The markets for nonperishable commodities probably follow other rules. Taking a closer look at, for example, the market for rice in Bangladesh, we can see that there are often aspects of food production that violate the idea of an open market. Where there is the possibility to store the product and where the extension of credit enters the picture, it is more difficult to establish the kinds of markets seen in the fishing example (Rahman 2007). The power relations that disrupt the market can be extensive and insidious. Thinking for a moment about the production and sale of rice, Rahman describes how the price swings according to the cycle of the harvest. Thus, the further one moves from harvest time the more scarce the rice is and the higher the price. Poor farmers, with limited production and storage possibilities, and pressing needs to pay back loans for seeds and equipment, often need to sell their rice immediately upon harvesting when the price is at its lowest. Those farmers who are able to wait until after the supply drops and the prices rise do better economically. According to Rahman, the terms of the loans often include fixed time limits, meaning that the farmer needs to pay the creditor immediately after harvest. Not to do so would mean that the farmer would not be able to borrow money with which to buy seeds for the next crop. Rahman writes, "In contrast to the 'ideal-type' markets used in economic literature, 'real markets' are very diverse and complicated, having widely different formal and informal institutions, and economic context, particularly in developing countries like Bangladesh" (2007, 1). In this case, getting information on the spot price of rice via the mobile phone would not have the same effect as it did for Jensen's fishermen.

In spite of these limitations, the use of the mobile phone changes the flow of information in some markets. This has made the markets more efficient, reduced waste, and allowed both buyers and sellers a more stable economic life. In a real way, the mobile phone has changed the social ecology of this production process.

Once the farmer or the fisherman has cash in hand, another set of issues arises, namely, the saving and transferring of payments. Here also, the mobile phone functions as a type of bank terminal or money-transfer machine for many people in the developing world. It has become an easy way to transfer money internationally and also within countries (Donner 2008; Kalil 2009). In particular, prepaid scratch-cards have, in some cases, become a type of currency that is used to transfer money over longer distances. People buy a top-up scratch card and send the number to someone to whom they owe money. The person receiving the card number can then either use it or cash it out at a store who will sell it to another customer. This type of transfer arrangement means that, for example, a husband working in the capital city who needs to send money home to his wife does not have to send money through the post, on a bus, or with another person traveling in that direction.[27] It is more secure and there is less chance of theft. In Zimbabwe, for example, this system is sometimes used to pay bribes at security checkpoints. There are also commercial money-transfer systems, the most well-known being the M-Pesa system in Kenya, Tanzania, and Afghanistan. This system allows the transfer of small amounts of cash between subscribers without the need to visit a bank. The mobile phone has allowed for a new logic of organization. Both in the case of the fishermen and in the case of money transfer, it has supplemented or replaced an existing system. In other words, it has changed the social landscape of commercial activity. As noted by the early-adopting Norwegian salespeople, the mobile phone saved them time and made their work more effective. These benefits have since spread to developing countries.

Conclusion

As mobile communication has diffused into society, it has rearranged the social ecology of mediated communication, just as the mechanical clock changed the way we coordinated interactions and the automobile changed the nature and dimensions of transportation. Like these other technologies, the mobile phone moved from being useful to becoming essential, and is now taken for granted. It has changed the interaction between parents and their children. It has changed the way we coordinate everyday life, and it has changed business and commerce around the world.

The fact that we can come into direct contact with one another, and the fact that we can both author and receive communications while performing other activities means that the mobile phone has rearranged the communications landscape. It has done this at the expense of alternative communication systems. The mobile phone is being used in place of landline telephony and has dramatically reduced the importance—and viability—of telephone booths. It has changed the norms of coordination, the way we think of safety, the interaction between parents and children, the practice of political action, and commerce. All of these have been touched by the diffusion of mobile communication. All of this comes from a technology that was generally unknown only two decades ago.

The mobile phone has reinvented the communications landscape. By individualizing communication we can be in contact with our nearest social sphere whenever and wherever we (and they) may be. Indeed, as mobile telephony has established itself in the social ecology, we expect this kind of access of one another. It is this reciprocal web of expectations that I will focus on in the next chapter.

8 "It Is Not Your Desire That Decides": The Reciprocal Expectations of Mobile Telephony

In many places, mobile telephony is perceived as having reached critical mass. We have arrived at justifications governing its use, and it has rearranged the social landscape. In short, the mobile phone has become an established part of society. We may grumble when others' phones disturb us, but on the whole we like the personal accessibility provided by the mobile phone. We feel the need to have one with us and will often return home if we forget to take it with us.

The mobile phone has found a stable position in our lives. We have found that it is easiest to carry it in a particular pocket or in the inner pouch of our purse or backpack. We know how much we are willing to spend on it, and we have developed routines for navigating the menu structure, writing texts, charging it, and cleaning it. We know that ever since we dropped our telephone in the sink, the button on the left needs to be pushed three or four times before it works. We usually place it on the desk behind the paperclip container when at work and on the table in the entry at home. We have developed tricks to help us remember to take it as we leave the house, and we feel a sense of loss if we suddenly remember that we don't have it.

Beyond our own personal practices, we know the routines of others. We know, for example, that our teenaged friend Dillon likes to text, Melanie likes to use Facebook updates on her phone, but Aunt Nancy would prefer that we call. We know when it is useful for coordination ("John, the meeting has been moved to room 35, I will be there in 10 minutes, Sally"), and we know the frustrations it can cause ("Hi Sally, I am sick today, it looks like you will have to do the presentation yourself, sorry, John"). It occupies a somewhat stable physical, temporal, cognitive, economic, and social position in our lives.

In previous chapters, we have seen the mobile phone move from being owned only by the technical or economic avant-garde to being a device that can be as mundane as wristwatches and glasses (even though they are mundane, it is unthinkable for some people to leave home without these items). As the mobile phone has moved through this progression, it has gone beyond being only a useful gadget that is nice to have to becoming a necessity. Beyond the many personal benefits we receive from the mobile phone, we also expect others to have one (Urry 2007, 176, 223). This is the key thought: The mobile phone is essentially a reciprocal technology. With it, we can reach others in real time, and indeed we increasingly expect this, just as others expect it of us. As Katz suggests, it is problematic when we (or they) do not meet these reciprocal expectations.

To understand the importance of the mobile phone, we have to look at it from the position of others. We have to understand their need to get in touch with us as well as their eventual frustration when we are not available. Increasingly, we place ourselves outside of the normal dealings of our group if we refuse to have one, or if we forget or ruin our own mobile phone. In such cases we are, to some greater or lesser degree, being irresponsible toward our group. This is perhaps most keenly obvious for close family, friends, and work colleagues. When there is a minor family emergency—a child falls and breaks an arm—the spouse who is dealing with the broken arm needs to be able to contact the other partner so that other children can be collected from activities and other ongoing activities can be dealt with. We also see that the mobile phone has gained a position as an essential link between people in the case of major catastrophes such as bombings and terror events (Dutton 2003; Katz and Rice 2002; Cohen and Lemish 2005; Sundsøy et al. forthcoming). To not be available to one another in a small or a major crisis makes the situation much more difficult to manage. This reflection also illustrates how not having a mobile phone is a social and not just an individual problem. Just as with someone who refuses to use a clock or who always needs to hitch a ride, someone who is not available via the mobile phone becomes a problem for the other people with whom they interact. It makes it awkward to include them or to rely on them in group activities. In those situations where mobile telephony has reached critical mass, a social imperative will demand its use. In the words of a US high school teen, Noah, interviewed in 2009: "[My parents] always needed us, they always needed me to have it, more than I needed to have it. Like, my mom always needed to reach me, so she would never take away the phone as punishment, because it was more useful for me to have it than for her to take it away." Noah's parents see it as a lifeline (Urry 2007, 178). The mobile phone has made us individually available just as it makes us responsible to our social sphere in a new way.

In addition to the enhancement of our social interaction, the mobile phone can provide us with flexibility and the ability to make and change agreements on the fly. However, ownership and use comes with a price. Having a mobile phone, and, more to the point, others knowing that we are available on it, creates a responsibility. This comes through in comments from a 2003 group interview:

Ingebjørg: I have a little example. A couple of weeks ago I was going to meet a person at Majorstua[1] who I did not know very well. I couldn't find her. But I knew she had a mobile telephone so I called her. She was just on the other side of the store I was in. If we hadn't had telephones then wouldn't have . . .
Marianne: Yeah, it is both good and bad. I do not think I like it this way, that we call one another all the time. It is such that we organize it that way and so we do it that way. There is a lot of that.
Ingebjørg: I think that there is both a freedom and a disadvantage. Sometimes it is nice to have the possibility to be impulsive instead of planning time and place or if we are not sure, we want to go out. Now I can go to a cafe with a friend and just call [and say] "Where are you? Are you ready?" If you agree on that before then you think that you have to go since you have agreed to it. *It is not your desire that decides.* (Emphasis added)

These people are grappling with the difference between having a mobile phone to fulfill their own individual needs versus having one because of responsibilities to their social sphere. Ingebjørg comments that "It is not your desire that decides." The impetus driving adoption of social mediation technologies goes beyond individual agency. If we do not have one we may experience some blessed silence and not be bothered by other's need to call or text. However, we will likely be urged and cajoled to buy one by others who need to contact us. The importance of being always accessible is seen in the comments of Grete, a divorced mother with two children aged 12 and 8:

Interviewer: How would it be to go a day without a mobile phone?
Grete: That wouldn't work. No. It is like you always need to be available, or you think you always need to be available.
Interviewer: Who are you thinking about?
Grete: The children. They need to be able to reach me.

Not being accessible is not an option for Grete. She feels the need to have direct access to her children, and vice versa. It is one of the important links through which they plan, assist, and care for one another. It is also there in case something unexpected happens.

The device collectively gives us efficiency and it gives us contact with our social sphere; however, there is also an individual price to pay. We have to buy a phone and we need to subscribe to a service. In addition, we need to maintain our phone and we need to remember to have it with us. So long as we are willing to pay that price, we are accessible. Otherwise, we are out of the loop. Perhaps more importantly, it is not completely our own choice as to whether we need a phone. It is not completely our own desire that is at work, but rather that of our social sphere.

Since we have mutual expectations of use, we increasingly use the mobile phone to adjust the way we coordinate interaction, contact others in our intimate sphere, work out the issues that populate our daily lives, and ultimately carry out the projects of developing and maintaining cohesion in the intimate sphere (Campbell and Russo 2003; Castells et al. 2007; Haddon 2004; Ling 2008) and reinforcing strong social ties (Igarashi, Takai, and Youhida 2005; Ishii 2006; Kim et al. 2006; Leena, Tomi, and Arja 2005; Ling and Stald 2010; Ling et al. 2003; Matsuda 2005; Miyata 2006; Reid and Reid 2004; Smoreda and Thomas 2001; Wei and Lo 2006). In the words of Christian Licoppe (2004), we are experiencing connected presence.[2] This style of interaction carries with it the expectation that members of our closest sphere be available to us and us to them.

The Transition in Our Understanding of the Mobile Phone

"I Would Never Buy One for Private Use": Utilitarian Individualism

It has not always been the case that we have viewed the mobile phone as a social technology. Earlier in its diffusion, before it reached critical mass, it had more the profile of a personal and perhaps utilitarian technology, rather than a social expectation. In 1995, it was seen as a type of tool. Nonusers had difficulty understanding what use it might have. So long as the individual did not have a specific personal need, there was no reason to invest in one. This is seen in a 1995 group interview in Norway:

Interviewer: You three without a mobile phone, after you have heard the others here, will you go out and buy one?
Elaine: No, I would never buy one for private use.
Johanna: [. . .] I don't have use for it.

"I don't have use for it" summarized the situation for many people at that time. Based on their experience, and what might be called an ego-based perspective, no purpose was served by the mobile phone. The idea did not exist that others might need to reach them when they were away from a landline phone. The reciprocal notion of individual addressability did not arise. For some, there was a sense that ownership of the device would be more trouble than it was worth. In addition, the speakers describe their well-entrenched relationship to the landline phone, which was commonly available and through which they had learned to arrange their lives:

Frank: An important question I ask myself, do I really need this telephone? That is the most important question. I won't spend 1,000 kroner for something that I do not need. It is clear that this is dependent on job situation. I am in an office and there I have a telephone all the time. And at home there is a telephone and that covers my needs. It takes me twenty minutes to come home, and I do not need to call home to say that I am on the way. Then I can call right from the office. I have read in the newspaper that they sell 6,500 telephones every day just before Christmas. Then I ask myself the question: who is it that is buying telephones?

As the domestication approach (see chapter 2) would suggest, these respondents did a good job of describing the social understanding that had been built up around landline telephony. At that point, the logic of communication assumed use of the landline system. The fixed device at the office and another at home covered the needs of these speakers. Frank had built up his routines and expectations based on the landline telephone. His activities were developed and coordinated based on the movement between these fixed poles, supplemented with the use of an occasional telephone booth. The sense of universal accessibility that would come with the mobile phone was not a part of the picture.

In some cases, the mobile phone was a way to bring telephony into new frontiers. In the mid-1990s, it was prohibitively expensive to have a landline phone at a cottage in the Norwegian forest. Stringing the wires to remote locations was not possible. Thus, the mobile phone offered the chance for the first time to provide service. This did not mean, however, that the participants had accepted the notion of a personal communication device. They still saw the mobile phone as a somewhat more flexible landline—that is, a shared—telephone. We can see this in the comments of David in 1995:

David: But perhaps you have a cheap used phone that you have at the cottage, the few times that you go to the cottage, it is there as security at the cottage. I think it is OK to have one that you can buy for 500 kroner or 1,000, and you can leave it at the cottage, and you can use it to call from there the few times I have a need for that, or if the kids at home need to call me or you can call and check on your elderly mother or father.

In spite of the fact that he was speaking of a mobile, or at least a portable, phone, David had not completely adopted the notion of a *mobile* phone that was personal. Rather, he seems to see it as a type of fixed telephone that does not happen to have wires (Agar 2003). To be fair, mobile phones at that time were only starting to develop beyond the often cumbersome devices that were not easily portable. The point remains, however, that there is no sense of individual addressability, only place-based access. David had, in a broad sense, the same telephonic paradigm as an elderly woman, Martina, in the same set of group interviews, who said, "I think it is fine that the telephone is stationary at home. A conversation should be between two persons." David and Martina had not made the paradigm transition from landline telephony, and the assumption that the phone is attached to a particular place. This is perhaps not that surprising, since, at that point, telephone use was generally located in a particular spot in the home, perhaps with a special chair that one could use while talking. The idea of a personal telephone that gave us individual addressability was yet to be developed. Given the assumption of place-based telephony, there was not the general expectation that we were always available to others. Just as in the 1400s, when clock time was not generalized and was only available to those within range of a clock tower, with landline telephony we were only available as long as we were within earshot of the telephone.

For those who had established their lifestyles based on the landline telephone, there was little interest in mobile telephony. This is seen in the comments of Marta in a 1999 focus group in Norway.

Marta: I have been opposed to mobile phones because I think that they are a little too much. But on my seventy-fifth birthday my children and grandchildren gave me one, not that I was so excited about it. I didn't understand what I would use it for.

Jan: But now you like it?

Marta: Not so much. But I have it when I am out and driving, and I have a cabin, so I thought that I could be at my cabin alone, because I think it is nice there. I was there one night and I didn't like being alone. I basically don't use it at all.

Jan: But that one night, did you feel secure having it then?

Marta: Yeah, you could say that.

Interviewer: When was the last time you used it?

Marta: I don't know, it was a long time ago. I turn it on sometimes, but there are so few who call, so I don't have much use for it. But it is security if something should happen.

Mobile phone diffusion had come further than in the 1995 sessions reported above, but Marta had lived her life without the mobile phone. None of her social circle—or seemingly her family—used the mobile phone to contact her. Had they called, the mobile phone would probably not be on. Her life was ordered around the landline phone, and so, for her, not much was to be gained by having the mobile phone. Both she and the people with whom she communicated understood this. She did not have the job or the social ties that would push her to use the mobile phone. Thus, the device was reduced to being a marginally useful security link. Indeed, it may have given more peace of mind to her children than to Marta. This functionality might have been of use if she had felt comfortable alone at her cabin, but the dearth of social life there only made the experience lonely. Marta's situation shows how the landline, and not the mobile phone, was central to her sense of mediated communication.

Another characteristic of this period is that it was personal need and not social expectation that drove the adoption of mobile phones. We can see this in the comments of William, who applied a type of personal necessity test. He said, "I will think about my need for it. . . . Many people who buy it don't really have the need. Where do you need it, and [do] you need it?" The focus in William's comments is on the needs of the individual. He framed his thoughts in terms of "my need" and whether another person might need one.

While a strong individualist perspective prevailed among focus group participants, others began to think about the potential for individual access. This comes out in comments from the same 1995 focus group:

Interviewer: If we look forward five years in time, who do you think, in Norway, who will not use a mobile telephone?

Erling: My wife won't.

Interviewer: Why is that?

Erling: I can say this . . . if you don't have use for it, [people like my wife] are realistic enough to see that they will not buy one. But there are other motivations that are stronger, but people who don't have those motivations will neither buy nor use one. They are a little more modest about things than others. But there are fewer and fewer of them.

Peter: Yeah, but if she had gotten into an emergency, do you think that she would have called?

Erling: I don't know.

Clearly, Erling's estimation of his wife's situation was that landline telephony would likely fulfill her needs. He was not able to see why his wife might eventually need a mobile telephone. He was probably correct in noting that for some people in society, including his wife, the landline telephone was enough. Further, he notes that she did not have the same motivations or interest in new technology. It is only when he was pushed to consider emergencies that he became somewhat unsure.

"My Dependency Came Afterward": The Individual Adoption of the Mobile Phone

In spite of the utilitarian approach noted by the early informants, a sense was emerging that mobile telephony was reaching critical mass. It was still contested moral territory. Many were not completely comfortable with its implications, but many were starting to see that it could be useful. One of the 1995 respondents, Arnold, said,

> I don't have a mobile phone; I could have had one for work but have avoided it. I am a little against it but I have accepted that it has come to be, and [I] will buy one for private use and I buy it actually just to have one. Nice to have, not just that, [I] travel a little and will have it in the car because of security. I travel with my job, I work in high school and right now I am the principal at school but before [. . .] I traveled a lot. It enslaves me, the mobile phone. When you were free and you sat in the car and it rang. Actually, I don't have anything against the mobile phone. I have accepted its ugliness.

Unlike Erling's wife, Arnold was showing a begrudging acceptance of the mobile phone. Owning a mobile phone had been optional up to that point. It had been OK not to have one. Increasingly, however, the mobile phone was becoming something that was reasonably expected.

A tension arose between the positive and negative consequences of the mobile phone. It provided security and it was a beguiling technical device. Arnold noted a certain fascination with the gadget. It would be "nice to have." There was also, however, the baggage of social responsibility (Cumiskey 2005). For Arnold, the social world mediated through the mobile phone appears as a threat. The mobile phone would reduce the freedom from his obligations he enjoyed while he was away from the office or the home. Others could reach—and disturb—him. He no longer had the option to withdraw from society, since it is always only a phone call away. He did not think of the mobile phone as something that could facilitate his social interaction. Arnold was stuck in a type of "on-the-one-hand, while-on-the-other-hand" type dialogue. It was a useful device, but it came at a cost.

Others also understood this tension. It was an essential tool, but it was also an intrusion. These apprehensions are obvious in the comments of Lynn, a veterinary student interviewed at the same time:

> I have to have it. I am not interested that people can call me on the phone regardless of where I am and whenever it is. Free time is a little sacred. When I am a veterinarian, I will probably be dependent on it and then it will be 24 hours a day.

Lynn was anticipating her professional needs, for the accessibility was to become a part of her work. The mobile phone would be a central tool in this context. Her comments, however, indicate that this would come at a personal price, namely, a loss of private time. Faced with the same issue, some of the interviewees in the 1995 focus groups had accepted and, to some degree, embraced the mobile phone:

Ivar: It was because of my work that I got it. There was a system where I paid for it myself and I got it refunded from work. There was nothing special with it. It was used a lot during the first week and then it went down a bit. It was work that decided . . . I did not want it myself. My dependency came afterward, but my job asked that I have it with me so that they could get in touch with me. If I could have chosen, I would not have had it, and now later on it has been very important for me. I won't say that it is the big experience, but it is OK to have it with me in case something happens, but I didn't use it that much.

Like Arnold and Lynn, Ivar was somewhat ambivalent about the mobile phone. He got his through his job. It is almost with a sigh that he agreed to first use the mobile phone. He says, "I didn't want it myself. . . . If I could have chosen I would not have had it." Indeed, it was seen as an intrusion. There is the palpable sense that when he first got it, it was an imposition. It is also possible to suggest that it did not really fit into his daily routines. As suggested by the domestication approach outlined in chapter 2, Ivar and those around him had a set of routines. Their day-to-day activities assumed certain customs and practices. The mobile phone was not a part of that world. Because of his job, however, it became increasingly integrated into his everyday activities, and "later on it has been very important for me." The adoption process was not driven by the Ivar's need to call others, but, as with Marta, it was the social structure around him that pushed for adoption. His work colleagues and perhaps his boss wanted to have contact with him. Ivar's job provided the motivation. Unlike Marta, he accepted the mobile phone, and it soon found a niche in his routines.

"She Is the Only Girl in the Class without a Mobile Phone and That Is Really a Drag": The Transition from the Individual to the Social

Looking at the material from the late 1990s with a somewhat broad view, we can say that the informants often focused on the instrumental use of the mobile phone. However, as we saw in Ivar's comments, a sense was beginning to emerge that it could be used in running the errands of our social affairs. The mobile phone allowed people to be in touch with the members of their inner social sphere. One respondent noted, "For me [owning a phone] is only so that I will be accessible for the people who want to talk with me." When asked about the positive aspects of the mobile phone, John said these were:

Freedom of movement, a sense of security, others may have older parents that you are worried about, et cetera. The mobile phone helps to create a new sense of freedom that we did not have before. You have to be there, not just at work but also with the family.

Rather than simply getting a mobile phone to fulfill his own needs, John recognized that it was important that others could call to him as needed. Regardless of his location, he could still be available to family and also to colleagues.

By 2003, the 1995 ethos that one should be called via the landline telephone was breaking down, and the idea of individual access was asserting itself. In a 2003 interview, a participant noted, "If I need to get in touch with my friend and I call home, I never catch her there. I always need to call her mobile. She always has that with her. If she does not take the call it is something exceptional." The logic of the landline phone was being replaced with the individual addressability of the mobile phone.

By 2003, ownership was nearly universal among teens, and almost all middle school students owned one. Indeed those teens who were without one occupied a special social position. In a 2003 focus groups for teens, it was possible to see the contrasts. Respondents had experienced the transition and could remember the earlier generation of mobile phones. They had seen the shifts in social acceptance.

Wenche: Now, in a way, the mobile has come and the boundary for who can have one is so much lower. In the beginning I associated the mobile with my uncle who had a mobile telephone in a briefcase that we had with us on vacation to the mountains it was really heavy and now suddenly there are kids in elementary school that have one and parents are saying "No! You can't have one until you are 10." [. . .] the mobile phone has been less expensive so that, and smaller so that the man in the street can afford one, so in some ways it is very cool and everybody has one. When I think about the first mobile phone that my uncle had, it was, I don't know how I will describe it, but it was a mobile phone and it was huge. Now they sell phones on the street for one kr. [approx. 15 cents] and there are more and more people who have one and there are only a few people that do not have one.

Lilly: Yeah, I see it with my cousin. She is in eighth grade and she is the only girl in the class without a mobile phone and that is really a drag. Everything like should be sent in text messages and it is really a drag to not be a part of that. It is just because she has principled parents that she does not have one. She is looked at as odd or excluded, not because there is anything wrong with her but because she doesn't have a mobile phone.

Lilly's comments are particularly poignant. They can be contrasted with Erling's, the man who said that his wife would not likely ever have a mobile phone, and with Marta's, the elderly woman who received a mobile phone on her seventy-fifth birthday. Erling's wife and Marta had constructed a life where the landline phone was appropriate and defined the limits of their telephonic needs. Lilly, and her cousin, have a very different perspective. Lilly describes how her cousin is excluded since she does not have a mobile phone. The contrast between Erling's wife and Marta on the one hand, and Lilly's cousin on the other, indicates that in the period that had transpired, the mobile phone had assumed a different position in their respective groups. While Erling's wife and Marta could easily carry out their daily routines without a mobile phone, this was not the case for Lilly's cousin. For her, the assumption of individual addressability came into play. It was not only Lilly's cousin who had this assumption, but those in her class with whom she interacted. Unlike Erling's wife or Marta, Lilly needed a mobile phone to fully participate in her social circle. Not to have one placed her outside the flow of social events. These contrasts illustrate the shift in attitude that had taken place in less than a decade.

In the material from 1995, the idea often arose that the mobile phone was for instrumental, need-based use. This attitude softened in the next decade. We can see this in the unsuspecting attitude of a teen girl from the 2003 groups who was quite open in her willingness to use the device:

> Martine: I talk with a lot [on my mobile phone]. Sometimes someone calls the wrong number and so I talk with them also. I think that it is fun when someone calls the wrong number. And then I talk with three or four family members and friends and some people at school and things like that.

Martine's comments clearly contrast with those of William in 1995, who said, "I will think about my need for it. If you want to be reached all the time, you should have it. Many people who buy it don't really have the need. Where do you need it and [do] you need it?" Clearly, a sea change had occurred. To be sure, the mobile phone was still an important work-related medium for instrumental communication. However, it was increasingly seen as a way to casually contact members of our social network and, sometimes, even people outside of our normal group.

The progression traced here is that the mobile phone was first adopted by businesspeople and then began to be used for social purposes. For those who have adopted the mobile phone more recently, the progression can indeed go in the opposite direction. Cara Wallis documents how young "internal migrant" women in China (women who leave their home village to seek work in larger cities) often first get their mobile phone to maintain contact with family in their home village:

> For example, one woman I knew who had purchased a very basic phone to keep in touch with friends and family found that over time her mobile phone primarily served work purposes even as she continued to foot the entire monthly bill. As much as she said she valued her phone for sociality, she was the only woman with whom

> I spoke who said she would be relieved—it would "save her some worry" (*bijiao shengxin*)—if she did not have a phone since at work it had become a means for colleagues and bosses to constantly page her. Nevertheless, she could not get rid of her phone even if she wanted to. Her colleagues enjoyed her constant availability and her employer expected her to be always available. (Wallis 2011)

Wallis's informant had, in essence, started her relationship with the mobile phone in a later part of the cycle. The social dimension came first, followed by the commercial. A type of coercion was associated with her use of the mobile phone for work purposes. The mobile phone started as being a useful social mediation tool that increasingly became essential, indeed to the point of being a nuisance. However, the expectation grew that she would have the phone and use it for work. She was, in a sense, not allowed to be without a mobile phone. The weight of the social expectations was too heavy to allow this.

Thus, looking back over the period between 1995 and the present, we have seen the transition of the mobile phone from a rather strict sense of its being a tool to its being a device that is increasingly integrated into our social lives. Further, it is not only a personal tool; it is a technology of social mediation. The Faustian part of the deal is that many of the annoyances of modern life can reach us via the mobile phone. For some it is paradise, for others it is the abyss; and for some it is both at once. All of this fits neatly in our pocket. It is becoming taken for granted (Giddens 1986, 143; Chayko 2008, 123–126). Along with the internet, the mobile phone is becoming an assumed part of everyday life.

"A Weird Sense of Hope": The Mobile Phone as a Social Tool

In her analysis of mobile telephony among college students, Rhonda McEwen cited an informant, Jessica, as saying "[the mobile phone] gives me a weird sense of hope that people want to call me" (McEwen 2009, 132). This is what Wurtzel and Turner (1977, 256) called imminent connectedness. It is not necessarily our actual contact with others, but rather the potential for contact that is important. We carry with us the hope that others will call.

To facilitate that, we need to make ourselves available. We make sure the phone is charged, and we maintain our subscription. When it rings, we rush to answer the phone, in order to actualize the imminent connection. As often as not, these "others" are not unknown callers, but most likely those from within our immediate sphere. They are children wanting to know where their ice skates or their math book might be. They are friends with whom we are planning a dinner party, or they are spouses calling to rework the evening's obligations. Following the theme being developed here, imminent connectedness is reciprocal. Since we feel that we need to be available telephonically, we assume that others feel the same. Our mental construction of the mobile phone is bidirectional. Our use of the mobile phone will show us as social creatures, and we wish to receive the same treatment in return.

"The Boringest Week of My Life": Thwarting Impending Interaction by Being Unavailable

As we have seen, one way to clarify the social role of the mobile phone is to consider what happens when it is no longer available. An example of this comes from the history of the landline telephone. In January 1975, a fire at a switching center in New York City left a large portion of Manhattan without telephone service. All told, 144,755 telephones fell silent for 23 days. According to Wurtzel and Turner (1977), who studied the social consequences of this natural experiment, people coped with the loss by making calls from their workplaces or by using temporary public phones that were set up on the street. However, the respondents felt that not having a phone resulted in a loss of others' accessibility to them. This was felt perhaps even more keenly than the respondents' loss of the ability to call others. In other words, people noted both the inability to call others but also the inability of others to call them. Wurtzel and Turner note that "In sum, ubiquitous feelings of lost control suggest that loss of telephone contact is an assault upon the way sample members conceived and structured their social reality" (Wurtzel and Turner 1977, 257).

Compared with the landline system, the mobile phone allows for faster interaction, so that the device is even more thoroughly woven into our daily lives. Many people feel a sense of immediacy with the mobile phone, a sense that is felt even more keenly for its absence when we are, for some reason, "off net." According to Jenni, a Norwegian teen interviewed in 2003, "For me it is like when I send an SMS I expect to get an answer immediately. If you write an email you think that it can take all day, not always, but you think that you get the answer during the day." Sheila, another teen interviewed in the same group, said, "I have noticed that when friends send an SMS to me, they expect an immediate response." The sense that we—and others—should be immediately available and able to either answer voice calls or respond to text messages plays on the idea that we are more-or-less continually connected. There is a common expectation that we can always quickly respond to calls and texts. If we're not available, the situation becomes awkward. The loss of our mobile phone, for example, disturbs patterns that we have built up. It disrupts the flow of work and social interaction. This was seen in the 1995 comments of Bjørn, who used the phone for work purposes:

> I felt very lost without [the mobile phone] once I was used to having it, that people could get in touch with me all the time. I could get in touch with people when I needed to. It was out of service for two days when I drove to Kongsberg, Drammen, and Hønefoss, all of eastern Norway more or less. I felt lost without it. I had to stop at phone booths . . . It was stressful.

Bjørn describes the difficulty of coordinating his work as he drove between cities in southeastern Norway. The trip between these cities could take up to almost two hours if traffic was flowing. Were an appointment to be canceled, it would mean that he would waste half a day driving back and forth. In his situation, the need for coordination is obvious.

Another dimension of losing a mobile phone is that it has become a repository for personal information. We see this in the comments of Ayesha, a teen interviewed in the United States in 2009, who had unintentionally ruined her phone:

> I was washing the dishes and talking to my friend. And [my phone] fell in the water, so, you know, I dropped everything and got my phone. And like I hurried up and took my battery out, and dried it, so it would work. Because it worked last time, so I thought it would work. And I didn't write my numbers down or nothing; I don't have none of my numbers. So it was hard.

Ayesha's comments show that her phone was not only a conduit for communication, but a repository for information. She needed to go through the complex process of reconstructing her social contacts. Even after she eventually replaced her mobile phone, she could not be fully "on line," since she had lost information on how to contact her social network.

One boy, Tom, interviewed in 2009, called the week in which his phone was broken "the boringest week of my life." He continues, "The first day without my phone, like, I didn't text anybody. I felt like 'Where did all my friends go?' like I moved or something, because no one knew my house number. So I just sat there, and it was during summer break too!" Tom and his friends had come to rely on the logic of being available to one another via the mobile phone. When that link failed, no other immediately available alternatives presented themselves. Another interviewee said that when he was without his phone he simply stayed at home rather than taking the trouble of walking to his friend's house to see if he was home (Lenhart et al. 2010).

This role of the mobile phone is not lost on parents. The data from the 2009 Pew study (Hampton et al. 2009) on teens and mobile phones shows that slightly more than 60 percent of parents report having taken away their teens' phones as punishment. Nearly 70 percent of girls had received this type of punishment, according to the parents who were interviewed. At one level or another, the mobile phone has become such a linchpin of the social network that its elimination disrupts social interaction.

Perhaps not surprisingly, we invest our mobile phones with emotional meaning. We have seen that Sara said it is her "security blanket." Some report panic if they lose their phones, and parents restrict the use of the mobile phone as a type of punishment. Others have described what is perhaps an overdeveloped reliance on the mobile phone. Anna, a teen girl interviewed in the United States in 2009, said:

> Yeah, like when I lost my phone, I had just got it, too, and I was hysterical. I was convinced that when my mom left . . . well, we had just moved and I was there by myself. I was convinced that somebody was going to break in and I was going to get snatched up. But, it was just horrible, like I was hysterical for like a good two three hours.

The mobile phone became especially important to Anna when she had just moved to a new home, where she felt more vulnerable. A recently divorced woman, interviewed in Norway in 2003, "completely panicked" when she did not have her mobile phone. A divorce or a move may occasion a disproportionate dependence on the device.[3] For these two, the phone had perhaps become a venue for compensating for things they had lost or a link to protection. Their reactions also reveal a highly social dimension of mobile telephony. There was a very real sense that the link to others via the mobile phone had become central. When that connection became threatened by not having a phone, it was overwhelming for them.

When Others Are Out of the Loop

It is clear that when people forget or lose their mobile phone, they may feel that they are missing out on interaction. On the opposite side of the equation, we find the frustrations of those people who cannot get in touch with potential interlocutors who are without their mobile phone.

People can be incommunicado for different reasons. We have already heard the comments of Lilly, whose teenage cousin was not allowed to have a phone. Interviewees from 2000 also commented on how awkward it was to have friends without mobile phones. It was difficult to organize social interaction. In some cases, those who were outside the circle of mobile phone users lost their social currency, since including them in social interaction was problematic (Ling 2002).

Another, perhaps more critical case is that of people who are normally available via their mobile phone, but who, for one reason or another, are not currently online. Amparo Lasén (2011) describes the role of the mobile phone between romantic partners. As a part of her analysis, she describes how the mobile phone creates "the obligation of being accessible." At the simple practical level, it can mean that the couple is less agile in their planning and coordination. In addition, when one partner forgets the phone (this is usually the male, in Lasén's material) or does not answer text messages, it is a source of worry for the other partner. Unlike forgotten keys or pocketbooks, a forgotten mobile phone can suggest other motivations to a partner with a suspicious turn of mind. It can be interpreted as wanting to duck out of sight for some illicit purpose. In some cases, Lasén reports that the offending partner parries the worries of the other by saying the battery on their phone is dead, or that they simply forgot it in the car. In yet other cases, which underscore the reciprocal nature of the mobile telephony, she reports on people who went so far as to borrow someone else's phone so that they could tell their partner they forgot their own phone.[4]

"Calling Requires More Effort": The Social Production of Communication Practices

Another way to see the reflexivity of mobile telephony is by looking at how we construct the rules of its use. In general, people observe a common sense of courtesy when calling and texting. These are socially constructed rules. We generally know when to call and how long to talk. We have a sense of when others are being genuine, facetious, hurried, or superficial. We adjust our communications based on the situation, and we assume that our interlocutors have approximately the same sense of calibration. In some cases, we get it wrong (this is the voice of authority speaking). We talk too long, or we make inappropriate comments. We send a text message when we should have called. In this give-and-take, we work out a common idea of what constitutes appropriate communication. It is the development of our expected stock of knowledge through social interaction (Berger and Luckmann 1967). These issues are perhaps most obvious among teens as they work out their maturing sense of how to use the mobile phone. Teen interviewees in 2009 talked about others who called and texted too much or were obnoxious in their use. This is seen in the comments of Drew in response to a question about the best and worst things about mobile phones:

Drew: Um, the best thing is probably just communicating with your friends when they are gone and the worst thing is how annoying people can get.

Interviewer: What do you mean, how annoying?

Drew: Like, when they are non-stop texting you and you are not texting them back and they just don't get it that you don't want to text back.

What Drew describes indicates that he and his friends have not completely worked out the rules for texting. The mobile phone is a device that quite literally reaches into our inner sphere and demands out attention. It can trump other engagements and assert itself. As Drew suggests, it can give us a channel to a cherished friend when he is away, but it can also be a nuisance. Because of this, we must necessarily arrive at a common sense of proper use.[5] These judgments operate at the level of the small group, but they can be generalized into a broader sense of appropriate use. Drew tries to indicate his reluctance to continue texting by not responding; however, the insistence of his texting partners becomes awkward for him.[6] In a somewhat broader sense, Drew and his friends are in the process of developing a common ethic of how to use the mobile phone, though his comments indicate that this work is not complete in the case of texting.

To practice the correct form of interaction, we need to understand the expectations of our communication partners. Is it better to, for example, call, send a text message, or use Facebook chat in a given situation? To make this decision correctly, we have to imagine how our interlocutor will receive and interpret the communication. In addition, we need to calibrate how much effort we want to invest in the communication. We see this in the somewhat inconclusive comments of a teen interviewed in the United States in 2009 in reference to texting:

Patty: You aren't as connected, you aren't as responsible [when texting]. Because having a phone conversation, especially with somebody who is not really close to you, sometimes I just feel that it takes more work, versus just texting. You have to try to keep the flow of a conversation, if there is silence, you know, you're trying to . . . Um . . .

Even though it is more demanding, talking is somehow better since it is more "connected." The decision to call or text involves an inner reflection with regard to Patty's closeness to the intended interlocutor. It also is determined by the nature of the communication and her sense of how much effort it will take her to keep the interaction flowing. This is weighed against the potential for being seen as rude by her communication partner if she chooses the wrong medium. The informants from the Danish focus group on mobile Facebook (described in chapter 7) were also in the process of working out these issues. The calibration of how to maintain contact means that they have to think about the nature of their friendship and how the person receiving the communication will react. One participant in the Danish group noted, "With my best friends I would definitely send them a happy birthday text message or even call them up depending on the time of the day. If it is acquaintances it would be a Facebook happy birthday and hopefully something a little more creative than the previous twenty posts." The simple act of sending a birthday greeting or using the mobile phone to either text or to call is inherently social. Patty's ethical ruminations, however inconclusive, indicate that she is trying to put herself into the role of her interlocutor. These deliberations about what is best also show that she is a social actor.[7]

Being available via the phone also involves having to calibrate the volume and form of our use. This sensibility is inherently social. Understanding this is essential in order to be a member of a community. Domestication theory would suggest that we are in a constant process of working out the role and position of these understandings. Taking this one step further, we expect others to employ the same sensibility. This is evidence that the mobile phone is not simply a tool for our own benefit, but a device used in the project of sociation. The very existence of these rules underscores the social nature the mobile phone. This can be seen in the comments of teens from the United States in 2009:

Pia: If [some of my friends] know that they have to talk about something that might be a little tough, they're in an argument with a parent or something like that, texting can be easier because you can think about how you want to respond, you are not just like on the spot on the phone when somebody drops some like big news and like "ah, ah, I don't know how to respond to this." Texting will give you some time.

Hal: You can delete things if you . . . like, I would, if it is like a big issue I always reread it and go "mmmm, maybe that is not the smartest thing to say." Whereas in a real-time conversation you can't go "oh, forget I said that."

Mona: It makes it more easier. Especially with personal situations that you are going though, like if your friend is like "what's wrong?" you know, it's like easier than talking on the phone.

Hanna: See, I would rather, if I'm like pissed off or something I would rather call my friends than text them about it. Because I'd rather hear them talking to me and being like "it's OK, everything is going to be fine" than have them say that . . . like read that on a screen, it is less personal.

Randy: Yeah, usually if there is some emotional crisis involved, I call them because it is just, it just doesn't have the same feel to it, you know, just text on a screen. Usually horribly formatted and completely ungrammatical text.

Pia: I think since calling requires more effort and more of your time, then it is taken better and more like a person caring for you than if it is just like a text. Because it is like "oh yeah, I'm like eating dinner and watching a TV show and oh, I am just going to say [indistinct]" and calling, it is like you take time out of your day to sit and talk to somebody.

Hal: That's why I like to text more because I feel like text is more "hey! I have a question or something, you know, get back to me when you can," whereas I feel a call is more "talk to me now."

Mona: Yeah, that's true.
Hal: Because you have to pick up right then and talk to them right then.

As these teens indicate in their discussion, there are many ways to interact via the mobile phone. The specific functionality used can depend as much on the nature of the interaction, the message to be delivered and the relationship of the interlocutors as it does on the technology that is available. On the one hand, texting can be strategic. The fact that it is asynchronous and has only limited "bandwidth" makes it useful when we need to "talk about something that might be a little tough." It allows us to edit our response and think through the best way to deal with the situation.[8] This can come at the price of being seen as inauthentic and lacking spontaneity. Regardless of how the communication is framed, it is based on a reflexive image of the other and their eventual response to the message or interaction.

We see similar discussions when considering the notion of courtesy vis-à-vis the mobile phone (Ling 1997, 2008). These discussions, however, are not only framed in terms of the relationship between the communicating partners, but rather are also in reference to the relation between the telephone user and the others who are copresent and within earshot. Etiquette has an implicit reflectivity (Toby 1952, 235; Duncan 1970, 69). Our observance of courtesy tells others how we would like to be treated in return. Duncan calls it a "dramatization of the self" (Duncan 1970, 266). Geertz suggests that courtesy is a reciprocally built barrier that both governs our interactions and buffers us when untoward events arise (Geertz 1972; see also Gullestad 1992, 165). Manners allow us to develop and maintain a common sense of the appropriateness of a situation in spite of the different threats that may arise (Goffman 1967, 65).

The mobile phone clearly represents a challenge to decorum with both those who are copresent and our mediated interlocutors. Because of this, we have had to develop a sense of courtesy with which to address these situations. In reference to copresent others, texting, ducking into less trafficked areas, and ignoring incoming calls are all strategies to deal with this.

When using the mobile phone, we take the position of the other, and this is a fundamentally social gesture. When considering our mediated interlocutors, we do this by thinking about the appropriate channel for communication and we do it by thinking about the frequency of our communications and their content. We also do it by creating and observing a mutual sense of courtesy in our use of the mobile phone. In all cases we consider what good form is, or at least what is adequate. We also expect the same from others. It is through these interactions that we develop a common sense of interaction practices.

The Evolving Reification of Mobile Telephony

We, in general, feel comfortable using mobile phones. They are increasingly interwoven into our daily lives and our expectations of one another. According to John Urry:

> One consequence of these various developments is that mobile phones (and increasingly blackberries and communicators) are not "extravagant" and "frivolous" but "necessary evils," naturally interwoven with the human body and always at-hand so making a mobile, communicating life (just) possible. So when people misplace their mobiles they are "lost," physically disabled because they have had removed their "natural" ability to talk with absent others, and socially, because they are disconnected from their networks. . . . A landline connection cannot satisfactorily substitute a "lost" mobile phone. People are lost in a no-man's land of nonconnectivity; without tools for coordination they would experience less corporal travel and reduced face-to-face encounters. (Urry 2007, 178)

This attitude toward mobile telephony is still emerging. Timekeeping and the car/suburban complex are in different positions vis-à-vis their social facticity. In the words of Berger and Luckmann, all three technologies have a form of reification:

> Reification is the apprehension of human phenomena as if they were things, that is, in non-human or possibly supra-human terms. Another way of saying this is that reification is the apprehension of the products of human activity as if they were something else than human products—such as facts of nature, results of cosmic laws, or manifestations of divine will. Reification implies that man is capable of forgetting his own authorship of the human world, and, further, that the dialectic between man, the producer, and his products is lost to consciousness. The reified world is, by definition, a dehumanized world. It is experienced by man as a strange facticity, an *opus alienum* over which he has no control rather than as the *opus proprium* of his own activity. . . . The objectivity of the social world means that it confronts man as something outside of himself. The decisive question is whether he still retains the awareness that, however objectivated, the social world was made by men—and therefore can be remade by them. In other words, reification can be described as the extreme step in the process of objectification, whereby the objectivated world loses its comprehensibility as a human enterprise and becomes fixated as non-human non-humanizable, inert facticity. (Berger and Luckmann 1967, 89)

The clock is perhaps the most reified of the technologies discussed here. Time and timekeeping are, for all practical purposes, taken for granted. We have left behind many of the struggles and discussions associated with the implementation of timekeeping and consider much of this apparatus to be a simple part of the natural order. The car has a less secure position. For those who live in the suburbs, the "inert facticity" of the car is seen as a relatively new wrinkle in history. Many have parents or grandparents who lived in traditional small towns or in densely settled cities, not the suburbs. Others know people who live in city cores where there is a well-developed public transportation system. While the car-based structure has become entrenched for many suburbanites, the origins of this transition are not yet obscured by the same span of time that separates us from the development of mechanical timekeeping. The mobile phone is a mere stripling in this context. There is not the same level of reification. Weber's iron cage has not formed as completely for the mobile phone as it has for the car, and it has not formed as completely for the car as it has for the mechanical clock. That said, we increasingly expect others to be available via the phone, and we have our phones with us to answer their expectations. Not being available to those in our immediate social sphere is becoming less tenable. As said by an interviewee, Martin, "You get a lot of abuse if you don't have a mobile phone" (interviewed in Norway in 2003). Referring back to Durkheim's social facts and their coercive presence, they are felt most directly if we choose to struggle against them (Durkheim 1938, 53). The "abuse" that Martin reports has some of this characteristic. The mobile phone has progressed beyond being a transitory technology that, in the words of Weber, is a garment that "could be thrown aside at any moment" (Weber 2002, 181). To be sure, as suggested by the domestication approach, we are still working out the details (Haddon 2003). We are deciding on what is good, or at least, what is appropriate use. In all of this, we apply reflexivity. This is happening not just because of the characteristics of the technology, but because it has become a conduit through which sociation (calls, texts, and Facebook updates) takes place.

For many (but not all) people, there is an implicit acceptance of the mobile phone. Billions of people have accepted the logic of mobile-based social interaction. We see it in the arrangement of our daily affairs, and we particularly see it when the system fails. My sister-in-law, with whom we began, felt the effects of falling outside today's mobile logic. When she forgot her phone on that Saturday morning she was out of the loop. She made it to the exhibition, but she was not able to contact her friends, and she missed her lunch appointment. By itself, the inconvenience was quickly forgotten and no lasting harm was done. Nonetheless, an opportunity for social engagement was lost. The episode illustrates how the mobile phone is in the process of becoming a technology of social mediation and how it is becoming taken for granted.

9 Digital Gemeinschaft in the Era of Cars, Clocks, and Mobile Phones

The Car, the Clock, and the Phone as Social Mediation Technologies

Clocks, cars, and phones are three very different types of technologies. The mechanical clock has been around for a long time and is thoroughly interwoven into society. The car was developed in the late 1800s and eventually entered into a symbiotic relationship with the suburbs. The mobile phone has only been commonly available in many places since 2000. In addition to having different legacies, these three technologies also cover different areas of our lives. The clock is for timing and coordination, the car for transport, and the phone is for talking, texting, and, for the well-heeled of the world, net-surfing. Seen through another lens, though, the three have much in common. Each is what I call a technology of social mediation. Each has diffused and reached critical mass in at least some portion of society. We have seen how legitimation structures have been used to justify them and how they have changed the social ecology. Finally, we have seen how we have developed reciprocal expectations of each other's use of them.

The mechanical clock, the personal automobile, and the mobile phone have each, to one degree or another, been diffused into the very fabric of society. Clock time began to be used in the 1400s, and became particularly widespread after the horological revolution in the mid- to late 1600s. As clocks became more reliable and portable, they became the standard for coordination of social activities.

Use of the car moved into society much more quickly. Its use was becoming established in North America, Europe, and Oceania before World War II. After the war, several factors came together that led to the flourishing of the car and the suburbs. Finally, the mobile phone became widely diffused in most countries of the world starting in the late 1990s. The device was adopted in industrialized countries during that decade, and then, based on inexpensive handsets and prepaid subscriptions, it was adopted by billions of people around the globe after 2000. Thus, each of the technologies is widely diffused. In contemporary society, we are unlikely to meet a person who does not use time and timekeeping. The personal automobile is nearly ubiquitous, and it is the rare corner of the earth where the mobile phone is not widely used.

Each of these technologies has also spawned legitimation structures (as well as counter-legitimations). In broad strokes, the legitimation of clocks and clock time is seen in the narratives that prize regularity in scheduling and in rules of etiquette that reward punctuality. Counter-legitimations include the stressful nature of always keeping track of time. Legitimations of the car note that it, perhaps along with a home in the suburbs, is seen as a status marker. Further, the car is seen as a symbol of freedom and speed. Counter-legitimations include discussions about how it pollutes and is dangerous. Finally, legitimations of the mobile phone revolve around safety, coordination, and social contact. We see it as a way to save time, to facilitate interaction, and to strengthen our connections with others. The counter-legitimations focus on the stress of being constantly available to others and on disrupting the social sphere with improper use. While the mobile phone is disruptive and can challenge our sense of decorum, we are nevertheless loath to not have it with us (Traugott et al. 2006).

These three technologies have, each in their own way, changed the social ecology. Mechanical timekeeping made other forms of time calculation obsolete. Clock time replaced alternative (and often idiosyncratic) systems of timekeeping with a universally available, abstract system that could be tailored to the needs of the individual, group, or person. The mechanical clock evolved from regulating the phasing of isolated social/cultural activities, specifically the cycle of prayer, to regulating goods production and other major social systems, including transport. The impact of the car on the social ecology is perhaps the most obvious of all. It has effected major changes to the structure of urban society. Indeed, the nature of cities and suburbs would be drastically different had the automobile never been developed. In the suburban landscape, where public transportation is minimal, the car is the de facto transportation system. The car, and the massive sociotechnical complex in which it is embedded, forces us to operate within its sphere. Its impacts include the complex system of roads, the development of shopping malls, the tailoring of commerce and social life to the constraints imposed by cars, and the expansion of cities.

In its turn, the mobile phone has changed how we coordinate our private lives, businesses, and social organizations, and it has changed the way our emergency services function. It has changed the social ecology of individual availability and informal interaction. Texting allows for text-based, asynchronous interaction that does not necessarily disturb other copresent activities, and the development of the mobile internet has enhanced our access to information and given us an expanded repertoire of social networking possibilities.

Finally, these three technologies give rise to reciprocal expectations of access and use. We expect one another to know the time and to be concerned with timekeeping. In addition, we reciprocally expect punctuality with varying degrees of precision. With automobile-based transport, we share the expectation that people are able to transport themselves to work, school, and social activities. These expectations are particularly strong in suburban areas where private cars are the only alternative. Finally, we experience social reciprocity through having a mobile phone. We live with the expectation that others will be available to us, and us to them. When we are without one, we have temporarily closed the door on our social circle. More fundamentally, we feel the awkwardness of forgetting our phones. We can have a sense of being lost and excluded from the flow of social interaction. We are also uncomfortable when others are not available via their phones. In short, we have developed a reciprocal sense of mobile communication. We can increasingly take for granted that our interaction partners also share these expectations of mobile communication. These elements have given the mobile phone a central place in our interactions. Using the Katz principle, we become a problem for our friends when we are not able to keep time, transport ourselves, or be available on our mobile phone.

The Synergies of the Clock, the Car, and the Mobile Phone

Beyond each having their separate trajectories, these three technologies are also linked with each other. The complex of transportation, coordination, and communication is central to contemporary society (Carey 1988). We have traditionally coordinated transportation using clock time. However, we are increasingly coordinating parts of transportation through iterative mobile-phone-based microcoordination.

As outlined in chapter 4, before the development of mechanical transportation, life was largely local. People did not move far from their home base. In this regime, time was not as essential in the coordination of far-flung activities. Before the development of electronic mediation, coordination of local affairs was done via word of mouth and with distant counterparts by carrying letters from place to place (Beniger 1986). These letters, of course, were nearly synonymous with physical transportation:

The speed of transportation was the speed of such communication. Though there were the odd examples of cannon-based messages or optical telegraphy, most daily communication was either face to face or based on written documents. The transition to electronic communication was a radical break from tradition (Giddens 1986, 123; Standage 1998).

> The most radical disjuncture of relevance in modern history (whose implications are far from being exhausted) is the separation of media of communication, by the development of electronic signaling, from the media of transportation, the latter always having involved, by some means or another, the mobility of the human body. Morse's invention of the electromagnetic telegraph marks as distinctive a transition in human cultural development as the wheel or any other technical innovation ever did. (Giddens 1986, 123)

This is not to say that transportation was out of the picture when it came to social impact. Indeed, the electronic communication revolution and the automobile revolution each changed the social landscape. The popularization of the car and the suburbs added to the sheer complexity of transportation. The coordination and movement of goods and people took on new dimensions. As the automobile freed us from the dictates of timetable-based public transportation, it has coerced us into extreme flexibility, to paraphrase John Urry (2007, 119). Before the mobile phone, this flexibility lacked a form of coordination. This, in turn, meant that we needed to more carefully organize the logistics of our interactions.

The ever-spreading suburban landscape meant that there were increasing distances we needed to cover in order to carry out our activities, but we had only a partially developed communication system.[1] According to Mimi Sheller and John Urry, the "multiple socialities, of family life, community, leisure, the pleasures of movement and so on, are interwoven through complex jugglings of time and space which car journeys both allow, but also necessitate" (Sheller and Urry 2006; Urry 2000, 59).[2]

Until recently, however, we had no system of communication capable of coordinating us while in transit. Before the mobile phone, communication terminals were geographically fixed. People who were moving about were effectively incommunicado. This was problematic in the suburban regions of home, work, business, and leisure, many of which were no longer located closely together. In the premobile communication era, the coordination of car-based transportation relied heavily on fixed coordinates in both geographical and temporal terms. We used absolute coordinates when agreeing on arrangements ("I will meet you at 11:30 at the corner of Pearl Street and Broadway outside Left Hand Books"). Alternatively, when we needed to make adjustments, we might have used a phone booth. It might also have been possible to leave a message, for example, at the restaurant where the meeting was to take place, but this depended on the forbearance of the people working at the restaurant and did not always work.

In other words, lack of direct, interpersonal communication channels meant an element of chance played a role in our arrangement of meetings in some distant corner of the suburban sprawl. Further, if plans changed or went awry while our intended meeting partner was also en route, there was no reliable way to alter the arrangements. Thus, while the car gave us personal transportation, it did not give us personal addressability. The growth of the suburbs without simultaneous real-time coordination meant that we built up practices with which to organize coordination, such as the reliance on concrete times and places to meet. This system, however, was inflexible. Until the development of the mobile phone, meetings needed to be carefully planned.

The mobile phone has allowed us much more flexibility when coordinating, or microcoordinating, activities (Green 2002; Ling 2002; Townsend 2000). Mobile communication has given us a direct interpersonal channel of communication that we can use to coordinate our social lives. We can be away from fixed landline telephones and also enmeshed in our remote social worlds. We can be traveling (hopefully not driving ourselves, since talking on the phone while driving is extremely dangerous [Strayer, Drews, and Crouch 2006; Hulme and Truch 2006]) but yet in real-time contact with our coterie. We can work out arrangements through a series of calls and texts. Indeed, the mobile phone allows for a spontaneous rearrangement of these elements should the coordinates for the meeting be in flux (Urry 2007).

At the same time, the mobile phone produces both more independence and more interdependence with others. At one turn, it gives us freedom to move about while maintaining contact; at the next turn, it ties us securely into the expectations of our social networks.

> Certainly, arguments about mobile work, flexible scheduling, changes in the duration and cycles of activities, proximity, distance, and presence might suggest that widespread social and cultural change in the practice and understanding of temporality is occurring. While the "speed" of modern urban life and potential fragmentation in social relationships via temporal changes can certainly be noted, mobile technologies also introduce opportunities for new continuities across space and time, previously disjoined through centralization. (Green 2002, 290)

The mobile phone allows us to efficiently coordinate transportation and, in turn, more effectively engage in social interaction. Being in transit used to be, in some ways, "dead time," where we were not able to be involved in social interaction. According to Scott Campbell and Nojin Kwak (2010), the mobile phone can add a new dimension to this time. As passengers, we can stay engaged with different parts of our lives even as we move from one location to another. We are able to call and plan our next move, or we can just touch base with our partner as we catch our breath before the next point in our daily rounds.

All of this challenges (or perhaps modifies) time as a technology of coordination. When there was no mobile-based communication, agreeing on a time along with a place was a nearly absolute coordination strategy. Mobile communication pushes aside time in the regime of transportation coordination, at least within the small group.

The mobile phone also changes car-based logistics, since it challenges (or perhaps supplements) time-oriented coordination, especially for smaller groups. The mobile phone allows iterative planning through the use of calls, texts, and, for some, social networking sites. However, this may not work for large groups. When more than a handful of people need to arrange or rearrange plans, it is more efficient to use more conventional types of time-based coordination.[3]

The car, the clock, and the mobile phone can each be seen as a social mediation technology. In addition, there are synergies between the three. Time has a long history of being used to coordinate transportation. Schedules are a central part of public transportation, and at a personal level we use time as well as location as a way to agree on meetings. The mobile phone has added some "give" to the stricter system of time and place. It helps us when we are late or when we need to rework an appointment. The direct interpersonal contact afforded by the mobile phone lets us either change the time of a meeting or adjust where it is to take place. Thus, although these technologies can be seen as separate, they can also be seen as part of the same coordination/transportation complex (Urry 2007).[4]

Digital Gemeinschaft

As we have seen, our ability to iteratively work out when and where we will meet up can involve mastery of the car, the clock, and the mobile phone. The complex of these three technologies gives us tools with which to tie together the members of our intimate sphere. We can reach across long distances to chat with a like-minded person and then arrange to meet and cultivate our shared perspective on things. We can console a partner or friend even if he or she is far across town, and in a similar way, we can share his or her joys in real time. The sum of these elements results in what we might call a digital gemeinschaft, playing on the suggestion by the sociologist Ferdinand Tönnies (1865–1936) and his notion of the contrast between community and society, or gemeinschaft and gesellschaft.

For Tönnies, gemeinschaft is associated with elements of kinship, neighborhood, and friendship, whereas gesellschaft describes self-interest, competition, and contractual arrangements. Gemeinschaft focuses on folkways and mores, while gesellschaft is focused on business, travel, and the sciences. Gemeinschaft is intimate and personal; gesellschaft is formal and legal. With gemeinschaft, there is by strong affectivity and group-oriented feelings; with gesellschaft there is a high level of individualistic calculation (Inglis 2009, 818). As with Durkheim and his contrast between mechanical and organic solidarity, Marx with the contrast between feudal and capitalistic production, and Weber's increasing rationalization, Tönnies was focused on the transition from community orientation to a more individualistic society. With the tools of gemeinschaft and gesellschaft, Tönnies provides ideal-type models that can be applied to different social situations. He suggests that gemeinschaft is slowly but inevitably being replaced by gesellschaft. That is, the focus on community and the group is increasingly and inevitably being replaced with commercial orientation and self-interest.

> Gesellschaft . . . is to be understood as a multitude of natural and artificial individuals, the wills, and the spheres of whom are in many relations with and to one another, and remain nevertheless independent of one another and devoid of mutual familiar relationships. This gives us the general description of "bourgeois society" or "exchange gesellschaft," the nature and the movements of which legislative economy attempts to understand; a condition in which, according to the expression of Adam Smith "Every man . . . becomes in some measure a merchant. . . ." (Tönnes 2002, 76)

The tendency toward gesellschaft in society carries with it the capacity to organize power and commerce across the globe. Indeed, it is possible to argue that digital interaction is the handmaiden of gesellschaft, with its ability to manage logistics and trade on a global basis. In many ways, this is the legacy of the control revolution as described by Beniger (1986), and the transnational mobilities of "road warriors," as described by Elliott and Urry (2010). Following Tönnies, the ideal-type merchant is free of any ties that limit the pursuit of earnings (Inglis 2009, 820). The spirit of gesellschaft will spread across society at the expense of gemeinschaft. Interestingly, Tönnies posits that the metropolis is the milieu in which gesellschaft best thrives. It is an environment in which the elements of self-interest, competition, contractual arrangements, and individualistic calculation are at their height (Tönnes 2002). It is when the village and the town develop into a city that the transition takes place from gemeinschaft to gesellschaft. Indeed, according to Inglis (2009, 825), the metropolis is gesellschaft while the village and the town retain some of the familiar elements of gemeinschaft.

Many people have asserted that the city and also the suburbs are the loci of impersonality and social facades. Simmel (1971 [1903]) describes how in the city the individual needs to filter impressions and contacts. Riesman et al. (2001 [1950]) note that it is in the city that, paradoxically, we feel the greatest loneliness. McPherson et al. (2006) have also examined the issue of isolation in contemporary society. Their work shows that an increasing number of persons say that they have no one with whom they can discuss important matters. While not agreeing with these points, Hampton and Wellman (2003) note three common lines with regard to the transition to urban society: it leads to (1) the weakening of interpersonal community, including reduced contact with kin and friends; (2) disengagement from local neighborhood ties; and (3) a reduction in public participation. One is left with an image of the isolated individual driving home from work or shopping to his or her suburban home located on a well-manicured cul-de-sac that is devoid of people. He or she opens the automatic garage door, parks the car, and then goes directly inside, with little chance to interact with neighbors.

In spite of this depressing picture, there is also the opportunity for the retention of gemeinschaft. While much discussion has focused on the impersonal nature of urban centers and the suburbs, there has also been discussion as to how we maintain familiar groups in the face of the impersonal and automobile-based suburban setting. Gans (1967, 234) notes that in spite of predictions, isolation did not emerge in the suburbs; rather, the suburbs allowed for the cultivation of familial ties in the nuclear family. Fischer (1976) found that the size of the metropolis allowed people to find "birds of a feather," who perhaps lived in a neighboring or even more distant suburb. These are, to use the phrase of Calhoun (1988), communities without propinquity. This meant that people could find an outlet for special interests or subcultures (Fischer 1976). For example, there are enough people to populate the local gay scene, the badminton club, the 1950s car club, or any number of other interest groups. In other work, Fischer (1982) finds that people receive various forms of support and contact in the suburban landscape.[5]

The evidence indicates that community (or, as Tönnies would have it, gemeinschaft) has persevered. Wellman notes that local interaction continues to be important to our functioning as individuals in society (Wellman 1999, 2002). According to Hampton and Wellman (2003), neighbors continue have an important role in our lives. They provide various goods and services, such as lending us the proverbial cup of sugar, feeding the cats, or gathering our newspapers when we are away for the weekend. They note, however, that social interaction is only rarely based on immediate propinquity. In the spirit of Calhoun, they note that "The typical network community in North America consists of a small number of densely-knit immediate kin and a larger number of sparsely-knit friends, neighbors, workmates, and extended kin. . . . Neighborly relationships remain important, but as a minority of ties within the overall network community" (Hampton and Wellman 2003, 230; see also Gordon and de Souza e Silva 2011, 110). Like Gans, Hampton and Wellman (2003) find that length of residency increases the local ties that an individual accumulates.

Thus, on the one hand, car-enhanced suburbanism is thought to increase isolation. This pushes us in the direction, as suggested by Tönnies, that gesellschaft society inevitably will trump gemeinschaft. On the other hand, the evidence seems to indicate that gemeinschaft is hanging in there. In the words of Hampton and Wellman, "Private and parochial life continues to be important, with kin providing a stable core of broadly supportive relations and neighbors providing immediate access to tangible goods and services" (2003, 278). Gemeinschaft is being pulled and pushed by the various developments in society. It has to change and adjust how it is practiced, but it seems to still be a viable concept.

Against the backdrop of dislocated but enduring gemeinschaft, what is the role of the mobile phone? The mobile phone is, on the one hand, an instrument of commerce, that is, a device in the service of gesellschaft. In chapter 5 we saw how its first users were often salespeople who used it to make their work more effective. In a similar vein, it is easy to argue that the clock and the car have the same status, namely, technologies that are used in the service of commerce. Clock time is an essential element in the coordination of complex manufacturing and the delivery of services, just as the car-based shopping center is founded on the idea that people will drive a certain distance to shop; and, on the other side of the equation, the stores are supplied using trucks that are part of the same transportation infrastructure.

But the mobile phone, as well as the clock and the car, can also be seen as a technology of the gemeinschaft world. We use the car to visit friends and family, and we use clock time and increasingly the mobile phone to coordinate these events. Further underscoring the role of the mobile phone in the gemeinschaft world, we communicate with a very small number of people (Ling, Bertel, and Sundsøy 2012). Analysis shows that for an average Norwegian, half of all texts sent and received in Norway go to fewer than five different people. In the case of young teens the number is slightly higher, and in the case of elderly people it is lower.[6] Moreover, we generally are in contact with those who are not too far distant. Indeed, distance matters in relation to the strength of telephonic links. In an analysis of approximately 26 million calls and texts, about 25 percent were within the same postal zone and 50 percent were within 14 kilometers (8.7 miles). Only 25 percent of the calls were between people who were more than 100 kilometers apart (62.1 miles). The strongest links are usually less than 10 kilometers (6 miles) away (Ling et al. 2011). Thus we tend to talk to a limited number of people who are physically nearby. That is, they are within the radius of a short drive.

These observations point to the fact that the mobile phone connects us to a relatively small social circle, both geographically and socially. It is true that we have "long tail" contacts who are more socially and geographically distant. Nonetheless, the majority of calls are to a limited number of people who live within a "city-sized" area. These numbers describe the resilience of gemeinschaft. Although the data used here do not give us insight into the nature of the ties—are they based on kinship, neighborhood, or friendship?—they do indicate that the circle is quite small. The contact list in a typical mobile phone is not akin to the unmanageable number of "friends" spread across the globe as you might find on a teen's Facebook account. Rather, there is a core group with whom we are in contact. These numbers come to life when talking with parents and teens. In the focus groups, people noted that the mobile phone was used to stay in touch with friends and family. Adults described being able to call their partners or other friends and family members, and teens liked the mobile phone because it gave them an easy way to be in touch with friends. This attitude is perhaps best summarized by Mike, a teen in the United States who said, "The best thing [about the mobile phone] is easily being connected with people."

Other researchers have examined this issue and found that the mobile phone increases cohesion in the intimate sphere. Kenichi Ishii (2006), for example, found that the mobile phone helps people to maintain existing social bonds. Christian Licoppe (2004) has suggested the same in his analysis of texting in France, namely, that the mobile phone serves as an active link through which people sustain social connections. In other words, gemeinschaft, digitally mediated through the mobile phone and supported by car-based transportation, is alive and well.

In the face of an ever-expanding suburban landscape and the hassle of daily life, in the face of forces that often tear at the social fabric such as the need to have two jobs and the stress of getting to our or our children's activities, the ability to coordinate or to simply touch base with others gives us the chance to ground ourselves. We also use it to coordinate the mundane affairs of daily planning and make our daily routines more effective. In addition, we can also contact a friend for a quick chat or to send him or her an offhand text full of inside references, just for the fun of it. This can be done anywhere and anytime, since we are calling and texting a person, not a place. Increasingly, we can also post a Facebook update or we can claim "mayorship" of our favorite cafe in foursquare. These can be done in the moments between activities and need not require any serious orchestration. To use the phrase of Ito and Okabe (2005), they are a simple, digital "tap on the shoulder" that let us know our friends are still there. The legacy of these functional and also more impulsive communications is to tie us more firmly together. It is these ties that provide the social group with resiliency as it deals with the tensions of quotidian life.

We use the mobile phone, along with other technologies of social mediation, to manage the intimate sphere and to maintain the resilience of gemeinschaft. Suburban life has changed the way that this is practiced. In a daily existence of jobs and other pursuits that pull us to the far corners of the city and suburbia, we are able to use these technologies to arrange our lives and keep in touch with our closest relations. The car may have led to the expansion of the metropolis, and the clock may have provided for complex scheduling, but the mobile phone makes the metropolis more intimate. This speaks to the power of sociation—the power of gemeinschaft. The fact that we choose to call and text one another even as the pressure of our appointments in this widespread landscape pulls us in other directions is evidence of our drive to be social. Seen in this way, it is not odd that we expect others to be available via the mobile phone. It is not odd that we have begun to take this for granted.

Notes

1 The Forgotten Mobile Phone

1. That is, a device with an open operating system, often a larger (480×800) touch screen, advanced network capabilities (HSDPA, GPRS, or LTE), and often a 1 GHz central processing unit (CPU).

2. In this book, I use the phrase "mechanical timekeeping" to refer to watches and clocks in general. It is clear that many of today's devices are more electronic than mechanical.

3. This pressure at the local level is described by Goffman (1963; 1967) as having to respect the ongoing nature of the situation. Goffman was speaking of face-to-face situations; however, there is also the sense here that not being in a position to receive a critical call or to transmit critical information (critical in the sense of the ongoing functioning of the group) is seen in the same way. We may feel that we are somehow letting down the group.

4. It is important to note that although many of these advanced phones are being sold, there is a much larger user base of traditional mobile phones with far more limited functionality. This is particularly the case if we think in global terms.

2 DeWitt Clinton's "Grand Salute" versus Technologies of Social Mediation

1. James Gleick (2011) describes how drummers in Africa were able to broadcast surprisingly detailed information using a similar system of audio links.

2. It is reported that the Rothschilds in London used passenger pigeons to communicate the outcome of the battle of Waterloo. Based on this advanced information, they were able realize a tidy profit on government bonds since they knew to buy before anyone else had news of the battle's outcome (Howe 2007, 695).

3. The fax machine has been around in different forms since the nineteenth century. However, it enjoyed its peak of popularity only in the 1980s and was soon replaced by email.

4. The notion of a population in the work of Rogers is somewhat imprecise. In one sense the population would be all the individuals in a country or all the individuals who have the opportunity to consume a particular good or service. On the other hand, the total population when considering, for example, iPod owners will, in all likelihood, never be every individual. Too many people will just never have an interest in buying one. In this case, the total population might be the eventual equilibrium number of persons who own the device. We will never really know that number since the introduction of new versions of the iPod, or an alternative music player, will mean that the counting goes back to zero before the last laggard has bought his or her version of the earlier device.

5. Durkheim also studied suicide as a social fact; see Durkheim 1951. According to his analysis, suicide, which appears to be one of the most individualistic of events, has a social dimension. By studying the rates of suicide in Catholic and Protestant communities, he found that the rates were lower in Catholic communities. This was, according to Durkheim, the result of differences in the levels of cohesion in the different communities. The analysis shows that not all behavior can be explained by psychological elements. Rather, even though suicide is always present in society, broader social forces are at work that shape behavior.

6. Thanks to Ralph Schroeder for helping me work out these ideas.

7. Critical mass is a concept from physics that describes the behavior of unstable material such as uranium. It provides a useful—but imperfect—metaphor for the diffusion of some technologies. In physics, the basic process is that at random moments, atoms spew off neutrons to the surrounding environment. If there are no neighboring atoms, the neutron will simply head off into the heavens without consequence. However, if a wayward neutron hits another atom of unstable material, it causes the second atom to also shoot out neutrons that in turn might or might not lurch into tertiary atoms, and so on. If we think of a few isolated atoms of, for example, uranium 235, they can shoot out neutrons without any further effect. The errant neutrons can be thought of as a teen learning to drive. Since the direction of their travel is sometimes random, the best strategy for the parent is to find a huge empty parking lot where no damage can be done. In a similar way, an isolated atom floating in the vastness of space does not have much chance of causing a lot of damage, since there are not any atoms nearby.

Critical mass is the amount of fissile material that is needed to sustain a chain reaction. As the density of atoms increases, the probability of the wayward neutrons hitting something else increases. Using the metaphor of the teen learning to drive, if there is not just one teen wobbling about in the parking lot but 10 or 100, the probably of a crash increases with each new car. At some point, there are so many cars that it would be difficult for an experienced driver, much less a teen, to move about. This is what happens as uranium approaches critical mass. There are an increasing number of neutrons in a more limited space, none of which breaks for passing atoms. Rather, they plow right in.

At the atomic level, each time there is a hit, the resulting impact gives off binding energy. With tighter and tighter clusters of atoms weaving about the atomic parking lot, there begins to be a very high probability that they will hit another unstable atom. Critical mass is thus a type of tipping point where the probability of neutrons hitting other atoms approaches certainty. A critical mass for uranium 235 is a mass of atoms about the size of a cantaloupe, for example. When a mass of uranium 235 atoms is assembled into an object of this size, there is a dramatic transition in the likelihood that the neutrons will hit something.

Biology also has a notion of critical mass. In some ways, this is the opposite of the atomic example just discussed. In the biological sense of the word, a critical mass is the number of a particular species needed to maintain the population in a particular location. If the number becomes too small, the individual members of the population may not have the opportunity to meet breeding partners and thus would not produce offspring at a rate that was in balance with the older members of the population dying off.

Claude Fischer (1976) uses the notion of critical mass when discussing his subcultural theory of urbanism. The basic idea is that cities provide for the development of different exotic subcultures since there are enough "birds of a feather" to develop the clubs and mechanisms that support the subcultures.

8. As noted earlier, there can be power dimensions at play in that the boss demands it of us.

9. As will be discussed below, the holdouts may create their own counterideology.

10. The social perception of critical mass can be seen in the adoption of different technologies. Early adopters of the telephone, for example, often bought the devices in pairs, one for the home and another for the office or the factory (Markus 1987, 505). After the development of switching, telephony started to gain utility from the network effect (Cherry 1977).

11. See http://www.technikum29.de/en/communication/fax.shtm.

12. An analysis I have done using material from Statistics Norway shows that 80 percent of those 16 to 19 years old reported using a social networking site in 2008 where only 5 percent of those between 55 and 67 reported the same (chi^2 (8) = 497.113, sig. < 0.001). This analysis is based on a random sample of 1,975 Norwegians aged 9 thru 79. Data were collected in four "waves" throughout the year to capture seasonal variation.

13. See, e.g., http://www.rense.com/general20/666.htm.

14. See http://www.youtube.com/watch?v=WAo_4-TYuZU.

15. Interestingly, the idea of evolution has also been applied to the development of technology (Arthur and Polak 2006). Using what Arthur and Polak (2006) call combinatorial evolution, they developed a computer simulation with a set of very simple circuits and let it combine in any number of random ways until they fulfilled a pre-

defined need. The general impression is similar to the idea of a million monkeys typing on a million typewriters. When two circuits were combined, they were tested against predefined needs that were a surrogate for the broader ecosystem. If they fulfilled the need, they were accepted and could continue their paired existence. If not, it was counted as a poor adaptation. After a quarter-million iterations, the simulation had created, among other things, circuitry that was the basis of a simple calculator (Arthur 2009.) Aside from the teleological illusions, their conclusion is that we do not make huge leaps in innovative development; rather, we slowly rearrange the building blocks we have in order to make something more useful out of them. The process with technology development is not simply the monkeys pounding away at the typewriters and producing the occasional work of Shakespeare; it is more purposeful than that. It is some person confronted with a need and then having access to different elements of a solution. The solution may be a blind alley, or it may be the step that allows the next person to make a new adaptation. There are those rare instances when, for example, a Faraday or a Curie has a flash of insight upon which others can build, but again, these flashes are based on solid preparation.

3 "My Idea of Heaven Is a Daily Routine": Coordination and the Development of Mechanical Timekeeping

1. For the purposes of this chapter, a mechanical clock will be a device powered by a falling weight, a spring, or (much later) an electrical battery or power cord. In addition, the device has some form of regulating the movement (an escapement) and an audio (e.g., a bell) and/or a visual (e.g., a clock face or a display) method of marking the passing of time. This definition covers the tower clocks of the medieval period, stationary clocks in the home, the chronometer, the quartz wristwatch, clocks on mobile phones, and the clock radio on our nightstand. All kinds of water clocks, candles, sand glasses, sundials, astrolabes, etc. are excluded, because they never gained the position of mechanical clocks. The ability to make a reliable device that gave people access to a common metric is the point. Sundials and the like might fulfill parts of this definition, but they do so only poorly.

2. The proscribed times for prayer in the Islamic cycle of Salah, which is timed according to the movement of the sun, retains some of these dimensions. For Sunni Muslims prayer should take place near dawn, at the solar noon, and after nightfall.

3. Astrolabes were developed by Islamic scholars to follow the phase of the moon (among a large number of other things).

4. The system of twenty-four hours per day, sixty minutes per hour was developed by the Egyptians and the Mesopotamians.

5. This does not mean that there were not efforts to do this. Orloj, an astronomical clock in Prague, has indicators for both types of hours.

6. According to Levine (1998, 56), the Chinese also used aromas to announce the time, by arranging different types of incense to be burned at different times of the day.

7. With the development of digital displays, many of these issues again arose.

8. Modern scientific clocks are now more regular than the spinning of the earth. Thus, the earth's rotation is out of kilter with the clocks, and not—as was the case with early clocks—the opposite. According to the most accurate clocks of today, over the period of a decade, the spinning of the earth can vary up to several milliseconds.

9. There are some stories of clock towers with clock faces only on three of the four sides of the tower, since the people in the fourth district were not willing to contribute to the purchase of the clock (Dohrn-van Rossum 1992).

10. Something of the same can be seen in modern society when a telephone exchange breaks down (Wurtzel and Turner 1977), or when the outage of a bridge disrupts traffic patterns (Chang and Nojima 2001).

11. The first portable clocks, usually table clocks, were developed using the fusee escapement in the mid-1400s. One of the first timekeeping devices intended to be worn on the person was Henlein's "Nuremberg egg" in the mid-1500s (Dohrn-van Rossum 1992, 121).

12. Wristwatches were available as early as 1790. They were, however, often decorative and feminine (Kahlert, Mühe, and Brunner 1986). The male version of the wristwatch had to await the turn of the twentieth century and the Boer War and First World War, when pocket watches were impractical given the demands, for example, of flying an airplane.

13. Before that, "punctual" referred to producing small pricks or punctures, issues relating to punctuation or spatial exactness. It was only in the early 1600s that it was used in reference to time.

14. In some situations, not being punctual is a problem only for the individual, such as when one is late for a bus or a plane.

15. These are a set of—what we see now as politically incorrect—figures representing vanity, greed, death, and finally an infidel. Goodness is represented by the Apostles who parade into view on the hour supported by the philosopher, astronomer, angel, and a chronicler, who round out the cast of the good guys. The clock itself has a variety of dials showing the phases of the moon, the zodiac, the angle of the sun, and a variety of other details.

16. A modern version of this is the Patek Philippe Calibre 89, a 1.1 kilo "pocket" watch with two faces, 24 hands, 1,728 parts, and 33 "complications" (functions), including the traditional watch functions, the time in a second time zone, moon

phase display, time of sunrise and sunset, and a thermometer. For those who are unable to orient themselves by other means, it also has displays indicating the century, decade, year, and leap year. Its price is $6 million.

17. Bells were also used to alert people of irregular events, such as the reading of proclamations, fires, attacks, and other emergencies.

18. Clearly, timekeeping could not mark irregular events such as fires, attacks, or irregular celebrations and rituals. Indeed, our contemporary use of bells and acoustic notification is associated with these types of events.

19. Not incidentally, time pressure and stress became a theme among humanist authors, a theme that has continued to this day (Eriksen 2001). This is not to say, however, that the pressure to keep track of time is felt the same way in all cultures, but it is, nonetheless, a theme. In some cases, power issues are worked out through the control of time. In other cases, there is an undue respect (or lack of respect) for existing time regimes (Levine 1998).

20. For their part, Glennie and Thrift (2009) suggest that Thompson overplayed the importance of industrialization in all this. They place the transition to the use of clock time earlier and note that there was widespread use of clock time from the 1600s.

21. Mumford (1963) makes a similar argument.

22. The Islamic cycle of prayer retains the use of the solar noon.

23. There are actually more that twenty-four. Some countries that straddle two or more time zones (such as Iran and India) compromise on a "half-hour" adjusted time zone in the name of national unity.

24. Chicago is located at 87° 36' west and thus is ahead of the center for the Central Time Zone.

25. However, it took Congress almost thirty-five years to codify what had become the de facto standard.

26. Indeed, Simmel (1971, 328; see also Elliott and Urry) suggested just this thought experiment when he considered that "if all the watches in Berlin suddenly went wrong in different ways even only as much as an hour, its entire economic and commercial life would be derailed for some time."

27. Indeed, there is enough confusion between Europeans who use the twenty-four-hour system and Americans who use the a.m./p.m. system. There are also other naming systems that underscore the potential for misunderstanding. In Norway, for example, "half seven" ("halv sju") means 6:30, i.e., half an hour *before* seven. For a Brit, "half seven" means 7:30, i.e., half an hour *past* seven. Even these seemingly small differences can cause problems.

28. In some cases, for example transcontinental calls, there is the need to take time zones into account; but this is still using the same system, only lagged in order to accommodate global placement.

29. See http://www.pairadicecruisers.org/.

30. See http://www.iceskating.co.za/.

31. See http://www.forsyth.cc/CES/family.aspx.

4 "Four-Wheeled Bugs with Detachable Brains": The Constraining Freedom of the Automobile

1. See also the discussion of the mobilities paradigm by Urry (2007; 1974, 18).

2. Thanks to Knut Holtan Sørensen for this line of thought.

3. To their credit, many other countries retained public funding for mass transit systems and more concentrated, collective forms of housing.

4. Most roads are a public good. There are, to be sure, privately owned toll roads. Although the car is a private purchase, roads are often a public works project. Because of this, there have been endless discussions as to how best to pay for their construction (Flink 2001; Urry 2007; Rose 1976).

5. Other names for what we call the car include the self-motor, locomotive car, autobat, autopher, diamote, autovic, self-propelled carriage, and locomotor. The name automobile was firmly associated with the device only in about 1899 (Bryson 1994, 197).

6. A similar "steam carriage" was built by Richard Dudgeon in New York in the 1860s (Rose 1976, 65).

7. In the second half of the nineteenth century, streetcars were the most central form of mechanized urban transportation. This allowed access to the city center from new satellite developments at the periphery of the city.

8. According to Urry (2007, 112), the development of the bicycle, at least in the UK, presaged the later pressure to develop an adequate network of roads.

9. Howe (2007, 224) notes that in the mid-1800s it took weeks for information to move from the East Coast to as far west as western Michigan.

10. It eventually cost approximately $400 (equivalent to $9,000 in 2009 currency) when the efficiencies of production were realized.

11. There were about 4 inhabitants per car in many European countries.

12. These are averages for the whole population. By 2006, the United States had increased automobile based transport to 25,000 km per person per year. The use of public transport was about the same.

13. Biking is also a viable form of transportation in Denmark. About one-third of all people in Copenhagen commute using bikes (Millard-Ball and Schipper 2010).

14. Barber has correctly pointed out that the automobile closed us off from direct social interaction with others on the street (Barber 2006; Hayden 2003).

15. Of course, troubling secondary issues emerge as well, e.g., the car is often seen as a waste of money, a source of pollution, a dangerous moving object that can kill people, and a venue for questionable dating behavior.

16. Fischer (1992, 27) feels the Lynds may have overplayed this in their analysis.

17. There is a very real sense that the landline telephone is bound to a particular place. Clearly, the mobile phone has an advantage in this way. It extends the ability to coordinate interaction when we are away from those fixed locations with a landline telephone.

18. The pressure for and against federal support for roads had been a part of the political landscape for generations. It was in the 1950s that the major development of the Interstate Highway System was realized (Howe 2007, 81; Snider and Sheals 2003). See also http://www.nps.gov/history/rt66/histsig/missouricontext.htm#_ftnref72.

19. Claude Fischer (2011) writes: "Federal financial backstopping, regulation, and housing subsidies directly made the home buying and selling business lucrative. More generally, the federal systems of retirement pensions, elderly healthcare, unemployment insurance, union protections, college loans, and the like benefited a large American middle class and made the purchase of private homes possible."

20. Indeed, moving further outward, the so-called exurbs or edge cities extend beyond the traditional suburbs and in some cases begin to develop their own identity as separate cities (Berube et al 2006; Hayden 2003; Bahr 2004).

21. The car has also overtaken some of the more personal aspects of our lives. Serious courting takes place in the car. Indeed, in 1967 a survey in the United States found that 40 percent of all marriage proposals had been made in an automobile—presumably in concert with other courting activities (Flink 2001, 162).

22. The mobility afforded by the car also meant that motels became the location of choice for illicit love affairs (Lewis 1983).

23. Drive-in theaters got their start in the 1930s, but have since waned in popularity.

24. Our reliance on the transportation system is often most clearly seen when it is disrupted. Chang and Nojima (2001), for example, have examined the impact of the Hyogo-Ken Nanbu earthquake. They estimate that about one quarter of the $6.5 million cost associated with the Northridge earthquake in Los Angeles was due to disruption of the transportation system.

25. This is not to say that these communities do not continue to thrive (Wellman and Haythornthwaite 2002).

26. Interestingly, automobile-based transportation has not completely replaced other forms of transport. In many cities around the world, there are viable alternatives or a different social structure (Schroeder 2007, 113). There are subways, buses, and, in some cases, animal-based transport. In Copenhagen, Beijing, and Amsterdam, it is possible to experience bicycle traffic jams.

27. The technologies of social mediation discussed here are not universal. This is perhaps seen in the discussion of another new technology, namely, the internet. It is instructive to examine the internet in the same way as has been done with timekeeping and the automobile and will be done with the mobile phone. Indeed, in many respects, some of the internet's functions have diffused into society and have had an impact on the social ecology. There are legitimation processes associated with the internet. Often we have the expectation that ourselves and others use the internet on a regular basis to, for example, check our mail or to check Facebook. Thus, as with the other systems considered here, the internet has a legitimate claim to being a social mediation technology. Like the automobile, however, some still live without it. For some groups, use is obligatory. However, at this point, that is not universally true. Nor is there is the same pressure to use it as, for example, there is to keep track of time.

5 "If I Didn't Have a Mobile Phone Then I Would Be Stuck": The Diffusion of Mobile Communication

1. Claude Fischer (1992, 43) describes farmers using barbwire fences in lieu of traditional copper wire when establishing early telephone networks.

2. A copy of the original memo describing the system can be found at http://www.privateline.com/archive/Ringcellreport1947.pdf.

3. It is interesting to note that, at about this time, Citizens' Band (CB) radio also enjoyed a short-lived popularity. During the late 1960s and early 1970s it became popular for cars to have a CB two-way radio.

4. A hertz is a single radio wave or cycle. The name comes from Heinrich Hertz, who first measured the length and velocity of electromagnetic waves.

5. The original Marconi wireless devices used an electric arc to generate radio waves. The spark that was generated used all the radio frequencies in the area covered by the signal. Since this system used Morse code for communication, and since there were so few people sending and receiving messages, this did not matter; it was either everything or nothing in terms of signal generation. When the operator pressed the key for a dot or a dash, the device used extensive parts of the radio spectrum. If two operators were sending in the same area, there could be confusion as to which signal

a listener should interpret. Eventually systems were developed that used only some of the frequency spectrum.

6. A yottahertz is one septillion waves per second, i.e., 10^{24} or 1,000,000,000,000,000,000,000,000 waves per second.

7. This is what you hold in your hand. Earlier versions often had a battery pack that could be the size of a small lunch box and a handset that you talked into. As the technology developed, the battery was been reduced in size and packaged into the handset.

8. The parts and materials for the mobile phone come from all over the world: tantalum from the Congo, nickel from Chile, the different plastic parts from oil pumped up from the Texas coast, Russia, Saudi Arabia, or the North Sea (Agar 2003).

9. For an excellent primer on the technical dimensions of the mobile phone system, see privateline.com, a site developed and maintained by Tom Farley at http://www.privateline.com/TelephoneHistory/History1.htm.

10. The data show that most of these telegrams were associated with congratulating a couple on their wedding, the birth of a child, or a similar occasion. About a quarter of them, however, were connected with work.

11. Indeed, the idea of using the mobile phone for business while on the move remains with us today (Urry 2007).

12. More advanced versions of pagers existed, including devices that allowed for the inclusion of voice messages. There were pagers with alphanumeric displays and two-way pagers where the individual could both receive and generate pages. Small-scale paging systems are still in existence, for example, in restaurants where customers are "paged" with a small device when their table is ready.

13. It is important to note that these are subscriptions and not necessarily people who have a mobile phone. In some cases, one person may have several subscriptions; in others, several people share the same subscription. Much of this depends on the economic situation of the user and the structure of the subscription system (Kalba 2008).

14. The similar numbers for Norway and the United States are 33 and 23, respectively. Also, as with other countries, there has been a drop in the number of landline subscriptions. Between 2004 and 2009 there was a reduction of 3.3 percent (ITU 2009; http://www.itu.int/ITU-D/ict/statistics/).

15. Unlike the system in the United States, where the subscription is often tied to a handset, the GSM subscription is tied to the SIM card, which can be switched from handset to handset as needed.

16. Another way to look at the role of mobile communication in society is to look at the rates of change in adoption. When comparing the number of subscriptions in

2009 to those of 2004, the countries that had grown the fastest were Nepal (130 percent increase in subscriptions), Tajikistan (105 percent), Guinea (105 percent), and Iraq (103 percent). Not surprisingly, the countries that grew the least were generally smaller countries who were early adopters of mobile communication. The slowest growing countries were Andorra (2 percent), Slovenia (3 percent), Taiwan (3 percent), and Norway (3 percent).

17. Rather than the number of subscriptions per 100 people as discussed above, the data in figure 5.3 look at the percentage of persons reporting that they owned a mobile phone. It is based on a series of surveys carried out on a random sample of the Norwegian population. These surveys were carried out by Telenor, the incumbent telecomm supplier in Norway. The author was the lead researcher.

18. The data contained in figure 5.3 come from surveys carried out by Telenor on the Norwegian population (in 1997, 1999, and 2001). Respondents were approximately 1,000 randomly chosen teens aged 13 to 20.

19. Texting allowed teens to develop a sense that they were establishing teen culture.

20. The material comes from a representative sample of 1,000 Norwegian teens carried out in the spring of 1998. The analysis was carried out by the author.

21. In the fall of 1997, Scandinavian telecom operators first started charging for text messages.

22. Similar analyses have been made in the United States (Ito et al. 2010) and in Europe (Livingstone et al. 2011).

23. Throughout this chapter, this material comes from focus groups that were arranged and largely carried out by the author. The Norwegian material has been translated to English by the author.

24. It is clear that other forms of mediation had arrived since the data collection in 1997. Instant messaging, blogging, and news groups were all a part of the web. These are not reflected in the data shown here.

25. Social networking sites are not used as intensely among those who are older. The same data set as used in figure 5.5 showed that about 1 in 4 35-year-olds reports using SNS on a daily basis. By way of comparison, about 90 percent of 35-year-olds reported sending text messages. In this data the use of social networking sites was almost nonexistent among 60-year-olds, but about half of this age group reported using texting on a daily basis. This indicates that while social networking sites have a solid role in society, texting has become a de facto form of mediation.

26. This analysis is based on a snapshot of the phones in current use on the Telenor network in September 2011. This included approximately 120 million phones in ten countries.

27. When measured in terms of bytes, the internet generates a large portion of the traffic. However, when considering the "events," i.e., calls, texts, and clicking on links, the mobile phone is still a very social device.

6 "We Are Either Abused or Spoiled by It—It Is Difficult to Say": Constructing Legitimacy for the Mobile Phone

1. The focus groups quoted in this chapter were organized by Telenor. The author analyzed the resulting data. The original focus groups were held in Norwegian, and the author has done the translation.

2. It is important to note that the landline telephone experienced the same transition (Fischer 1992).

3. In another version of this story, a person is in the bar at an expensive cafe. He (again, it is almost always a man) speaks loudly into his phone about buying and selling stocks at high prices. The story goes that in the midst of a particularly excited discussion of prices, the telephone actually rings, revealing that he was faking the call.

4. See http://www.youtube.com/watch?v=EsYRQkmVifg (visited on May 20, 2010).

5. Analysis shows that, in general, we call only a limited number of different contacts and that most of our mobile communication traffic is focused on these few numbers. Based on an analysis of over 188 million interactions over a three-month period, 75 percent of all messages on average go to 15 contacts or fewer (Ling, Bertel, and Sundsøy, 2012). Indeed, half of all text messages go to only about five different texting partners and over half our calls go to only three different numbers (Ling and Sundsøy, forthcoming; Kovanen, Saramäki, and Kaski 2011). These considerations suggest that the mobile is an instrument of the intimate sphere. It is the device that we use to contact our closest friends and family (Hampton et al. 2009; Ling 2008).

6. There is a debate about the effects of electromagnetic radiation. It is difficult for me to comment on the content of the science. There is, however, a social dimension to this issue, in that believing one or the other side of the research will affect whether or not we feel comfortable using a mobile phone. Many—indeed the vast majority of people—have considered the issue but are willing to use the device. Others are more concerned and think that the radiation is potentially dangerous. Some people feel that they have an extreme sensitivity to this type of radiation and go so far as to use special clothing and coverings to protect themselves from its effects (Bergløff 2009).

7. See http://mythbustersresults.com/episode2.

8. This includes staged material on YouTube. A Japanese version of the popcorn story is at http://www.youtube.com/watch?v=K1rRvExSuCc&feature=related.

Another version, where the popcorn does not pop, is at http://www.youtube.com/watch?v=dhBZs4Yho80&feature=related. Other legends suggest that by typing in a particular code, you will be able to activate a type of reserve battery, and that, should you lock your car keys in the car, you can call home and have the person at home "click" a spare set of keys into their end of the line, thus sending the signal through the telephone system and unlocking the car.

9. Females—both parents and teens—were more inclined to agree with this, i.e., females feel more strongly than males that the mobile phone provides a sense of safety. The Chi^2 for the mothers vs. fathers was (1) = 50.572, sig. < 0.001, and for the teen girls vs. teen boys was (1) = 7.113, sig. = 0.008.

10. A slight variation on this theme is that some will fake a phone call in order to avoid talking to people they wish to avoid.

11. According to Fischer (1992), the landline telephone was often marketed, at least partially, for its potential to assist in emergencies. Marvin (1988, 89) describes, however, that although this potential was a part of the technology, it sometimes was not included in the routines and regulations of emergency services in its early days.

12. Indeed, individualization (Beck and Beck-Gernsheim 2002), and in particular the role of technology in individualization (Lash 2002), is one of the great trends in modern society. The mobile phone is one example of this process, but others include the personal automobile, the personal computer, and a seemingly endless number of personal entertainment devices.

7 Mobile Communication and Its Readjustment of the Social Ecology

1. Indeed, while subscriptions to mobile telephony have skyrocketed, subscriptions to landline telephony are actually dropping (ITU-D 2010).

2. It has been reported that there are no phone booths in Jordan and that major carriers in European countries have also discontinued this service (see http://www.cellular.co.za/news_2004/march/032704-payphones_suffer_from_cellphone.htm).

3. To some degree, this is comparing apples to oranges. A landline phone often serves several people in a household, while a mobile phone is often associated with a single individual, particularly in the developed world. Thus, in Norway, the average home has slightly more than two people. This means that a 1 percent drop in the number of landline phones would translate to a drop of about 2 percent of the population being served by that system (assuming that the drop in subscriptions came from private homes and not offices, etc.).

4. When asked, teens in the United States said that they often used mobile phones to coordinate their lives. Data from the Pew study of US youth show that 40 percent of teens do this least once a day. The data also show that as teens mature, they are more likely to use the mobile phone to coordinate social interaction. Female teens

seem to use their phones to coordinate activities more often than teen males (f (1, 1143) = 19.538, sig. < 0.001). The data show that 34 percent of teen males report using the mobile phone to coordinate on a daily basis and 47 percent of teen females report the same. The data for this analysis come from the Pew study on mobile communication (Lenhart et al. 2010).

5. Approximately half of teens and young adults surveyed had a smart phone from which they could access the mobile internet. In the 2011 focus group of 12 young adults examined here, only one did not have an internet-capable handset.

6. This focus group took place in Denmark in 2011. It included 12 young adults who were largely smartphone users.

7. In earlier work (Ling and Yttri 2002), I have referred to this more expressive interaction as "hyper-coordination."

8. This also included taxi drivers who were using the device to organize their work more effectively (Townsend 2000) and prostitutes who used it to coordinate their work (Fortunati 1998).

9. The internet has also been a tool in this area, providing the most diverse groups a forum where they can collect information, schedule events, pamphleteer (or perhaps blog), and so on (Hargittai, Gallo, and Kane 2008).

10. The actual number is usually less, since an individual can receive such a message from several different persons.

11. This functionality is also used for the organization of "flash mobs" (Rheingold 2002). Another application is "home alone" parties among teens. In this case, a teen may want to have a party for his or her closest friends. However, if word gets out that there is a party, the message quickly spreads through the tree structure. In one case, 500 teens converged and proceeded to destroy the home: see http://www.nettavisen.no/utenriks/article1542499.ece.

12. The coordination of protests via the mobile phone has been an element in the change of governments. When there is a broad-based focus among the public and a wavering authority structure, the mobile phone has been a decisive tool in the hands of the protesters. However, when the authority structure is unified in purpose and there are disparate challenging coalitions, the logistics provided by the mobile phone are often not enough to take the day. Indeed, Gamson (1975) has commented on how fractionalization is often the bane of protest movements. The individual addressability afforded by the mobile phone means that it can be used as a tactical tool to coordinate social protests. It can be used to mobilize the public and to direct protest actions as they are taking place.

13. They are also called "flashes" or "beeps."

14. In some cases, missed calls are simply the result of the person being called not reaching the phone in time. However, the vast majority of these are intentional missed calls.

15. There are courtesies associated with the practice, e.g., the frequency of use, the promptness of responses, etc.

16. See, e.g., http://www.textually.org/textually/archives/2010/01/025300.htm; http://www.technologyreview.com/blog/mimssbits/26513/?p1=A4.

17. See http://www.textually.org/textually/archives/2007/04/015626.htm.

18. See http://www.textually.org/textually/archives/2005/01/006583.

19. See http://www.textually.org/textually/archives/2009/06/023931.htm.

20. See http://www.textually.org/textually/archives/2006/12/014261.htm.

21. It is often suggested that hikers should have a mobile phone with them. The mobile phone, however, is not seen as a primary form of rescue equipment. Appropriate clothing, food and water, first aid equipment, and the good sense to not take unnecessary risks rank higher on the list when considering emergencies in the wilderness. The mobile phone can, however, help to organize a rescue should a situation arise. See, e.g., http://karenberger.suite101.com/ten-essentials-for-hiking-and-outdoor-survival-a95914.

22. This is based on analysis of 49,895 subscriptions in the Norwegian network in January 2010. Clearly, there is also a "long tail" of other numbers that receive texts. These include a raft of numbers that are only texted to on rare occasions, such as texts to commercial services, distant friends, and relatives.

23. Indeed, as noted in the previous chapter, one in six teens in the United States sends more than 200 text messages a day (Lenhart et al. 2007).

24. It is unknown how many different numbers are commonly texted to in the United States. Given the high volume of texting by teens, it may be a somewhat larger group than in Norway.

25. This is not to say that people cannot "switch modes." Just as with other forms of communication, interlocutors are able to write in a more formal style when there is a need for that, and they can also adopt a more relaxed writing style when that is appropriate (Nielsen 2006).

26. Divorced parents may also use this strategy when dealing with their ex-partner. In this case, text messages are used to keep a distance and to limit the range of emotion that is brought into the interaction (Hjorthol et al. 2007).

27. This system is not without its risks. The number being sent might not be legitimate. In addition, there is no real system to control this type of money transfer as there is, for example, with traditional bank transfers. Another dimension of this is that the numbers are not necessarily traceable. It has been reported that ransoms have been paid in Iraq by using a long series of scratch-card numbers. Corrupt officials use the system to accept bribes, and prostitutes have regular customers send

retainers to their mobile phones. See, e.g., http://www.economist.com/node/14870118?story_id=14870118&fsrc=rss, http://www.comviva.com/media/news_BBC.pdf, or http://www.kiwanja.net/database/article/article_interview_jan_chipchase.pdf.

8 "It Is Not Your Desire That Decides": The Reciprocal Expectations of Mobile Telephony

1. A section in Oslo with a major train and bus station.

2. We are careful in the way that we allow people into this charmed circle of friends. Rhonda McEwen (2009), for example, reports on the strategies and thoughts of college students when sharing their mobile phone number. When two persons of the same sex exchange numbers, the sharing is reported to be based on specific needs. When the sharing is between sexes and there is the sense by one of the partners that sharing numbers might be interpreted as having romantic undertones, more elaborate stratagems come into play.

3. Overdependence on the mobile phone is indeed an area of study (Park 2005).

4. According to the work of Lasén (2011), 6 percent of women and 7 percent of men phone their partners more than six times a day.

5. A similar line of thought is suggested by Miguel Sicart (2009) in the case of games and gaming. In this case, the other actors might also include the game creators and other players.

6. In voice interaction over the phone, we use pauses between turns, and sometimes overt comments ("Sorry, I've got to go") to gauge the length of a telephone call (Veach 1981). In the absence of explicit cues, as the pauses between turns increase in length, it is a sign that the topic of conversation is winding down and that we must either find a new topic, or conclude the conversation. Veach (1981) found that young adolescents have not always developed the social antenna to understand this convention.

7. Twitter and the "quasi-broadcast" portions of social networking sites do not have precisely the same characteristics, since there is not necessarily the assumption that all the people who read these messages will engage us in an ongoing interaction. Thus, in some respects these are not social interactions in the same way as the one-to-one interactions of mobile telephony and, for example, interpersonal email.

8. It is worth noting that people do not always take the time to edit their texts. Indeed, "drunken" texting gives people the ability to show their unfiltered and less than flattering side (Hollenbaugh and Ferris 2010).

9 Digital Gemeinschaft in the Era of Cars, Clocks, and Mobile Phones

1. It is clear that the car has been a driving force in the expansion of the city. This is not to say that the landline phone did not also have a hand in this development:

> The telephone allowed for the separation of office work and other parts of the manufacturing process. Production, shipping, marketing, financing and administration could be located in different parts of the city, or even in different portions of the globe. . . . In the dynamic circumstances of recent times, especially in countries of advanced economy, the telephone has increased this spatial division of labor and society: the home could be more distant from the workplace, the office of a firm from its plant, the consumer from his supplier. The telephone provides, when needed, quasi immediate verbal communication between all these interdependent units at a minimal cost. (Gottman 1977, 312)

Although the landline telephone facilitated the expansion of the city, it was not as well adapted to the type of coordination needed for individual, personalized transportation.

2. During this period, the need for more nuanced coordination was seen in the use of two-way radio to control the movement of different types of fleet operators (e.g., taxis, freight delivery and police). In these cases, two-way radio communication facilitated the management of the fleet of vehicles (Urry 2007; Manning 1996).

3. Obviously, other technologies are in the mix. For some groups, the PC-based internet is very much a technology of social mediation, since it affords person-to-person as well as one-to-many "quasi broadcast" mediation (to some degree, this is being brought to the mobile world [Ling and Sundsøy 2009]). For some groups, net use, or more precisely, Twitter or Facebook use, is tacitly assumed. These services are widespread (at least within particular groups), they have a legitimation apparatus, have changed the social ecology, and reciprocity is a part of their use. Seen this way, these services fit into the analysis presented here. This means that platforms like Twitter, Facebook, Skype, Viber, and MXIT (Walton and Donner 2009) are becoming a part of our social interaction repertoire.

4. It is also possible to suggest that maps and location information constitute another technology of social mediation. Recent developments in the use of global positioning and its spread through society give us some sense of this (see, e.g., Gordon and de Souza e Silva 2011).

5. In more rural settings, kin, neighbors, and religious organizations are more important, whereas in suburban and urban settings, friends, work colleagues, and various types of club acquaintances are more central.

6. Data also show that, in broad strokes, the same can be said of people in Malaysia, Thailand, and Bangladesh (Ling et al. forthcoming).

References

Abrahamson, E. 2003. Hear me now: Competition, regulation, and innovation in mobile telephony in the United States: 1945–1983. Doctoral dissertation, John Hopkins University.

Agar, Jon. 2003. *Constant Touch: A Global History of the Mobile Phone*. Cambridge: Icon Books.

Allard, S. W., R. M. Tolman, and D. Rosen. 2003. The geography of need: Spatial distribution of barriers to employment in metropolitan Detroit. *Policy Studies Journal: The Journal of the Policy Studies Organization* 31 (3):293–308.

Andrewes, William J. H. 2002. A chronicle of timekeeping. *Scientific American* 287 (3):58–67.

Ang, Peng Hwa, Shyam Tekwani, and Guozhen Wang. 2009. When the mobile phone is cut: Lessons from Nepal. Paper presented at the International Conference on Mobile Communication and Social Policy, Rutgers University, New Brunswick, New Jersey, October 9–11, 2009.

Arthur, W. Brian. 2009. Where Darwin doesn't fit . . . *New Scientist* 203 (2722):26–27.

Arthur, W. Brian, and Wolfgang Polak. 2006. The evolution of technology within a simple computer model. *Complexity* 11 (5):23–31.

Astley, W. G., and C. J. Fombrun. 1983. Collective strategy: Social ecology of organizational environments. *Academy of Management Review* 8 (4):576–587.

Bahr, H. M., C. Mitchell, X. Li, A. Walker, and K. Sucher. 2004. Trends in family space/time, conflict, and solidarity: Middletown 1924–1999. *City & Community* 3 (3):263–291.

Bakalis, Sandor, Muriel Abeln, and Enid Mante-Meijer. 1997. Adoption of mobile telephony in Europe. In *Communications on the Move: The Experience of Mobil Telephony in the 1990s*, ed. Leslie Haddon. Farsta: COST 248.

Bakardjieva, Maria. 2003. Virtual togetherness: An everyday-life perspective. *Media Culture & Society* 25:291–313.

Barber, Melvin W. 2006. Abandoned communities: The malignant social consequences of modern technology on communities. *Journal of Evolution and Technology* 15 (1):37–50.

Baron, Naomi, and Rich Ling. 2007. Emerging patterns of American mobile phone use: Electronically-mediated communication in transition. Paper presented at the Mobile Media 2007, Sydney, Australia, July 2–4, 2007.

Bastiansen, H. G. 2006. *Det Piper Og Synger Overalt: Mobiltelefonen I Norge Fra Ide Til Allemannseie*. Oslo: Norsk Telemuseum.

Bauman, Zygmunt. 2005. Durkheim's society revisited. In *The Cambridge Companion to Durkheim*, ed. Jeffrey C. Alexander and Philip Smith, 360–382. Cambridge: Cambridge University Press.

Beck, U., and E. Beck-Gernsheim. 2002. *Individualization: Institutionalized Individualism and Its Social and Political Consequences*. London: Sage.

Bedini, Silvio A., and Francis R. Maddison. 1966. Mechanical universe: The astrarium of Giovanni De' Dondi. *Transactions of the American Philosophical Society* 56 (5):1–69.

Belasco, W. 1983. Commercalized nostalgia. In *The Automobile and American Culture*, ed. D. L. Lewis and L. Goldstein, 105–122. Ann Arbor: University of Michigan Press.

Beniger, J. R. 1986. *The Control Revolution: Technological and Economic Origins of the Information Society*. Cambridge, MA: Harvard University Press.

Berger, P., and T. Luckmann. 1967. *The Social Construction of Reality: A Treatise in the Sociology of Knowledge*. New York: Anchor.

Bergløff, Charlotte B. 2009. Ensom Kamp Mot Strålene. *Aftenposten*, August 22.

Berker, T., M. Hartmann, Y. Punie, and K. Ward. 2006. *Domestication of Media and Technology*. Maidenhead: Open University Press.

Berube, Alan, Audrey Singer, Jill H. Wilson, and William H. Frey. 2006. *Finding Exurbia: America's Fast-Growing Communities at the Metropolitan Fringe*. Washington, DC: The Brookings Institution.

Best, Joel, and Gerald T. Horiuchi. 1985. The razor blade in the apple: The Social construction of urban legends. *Social Problems* 32 (5):488–499.

Bhuiyan, Serajul I. 2011. Social media and its effectiveness in the political reform movement in Egypt. *Middle East Media Educator* 1:14–20.

Bijker, W. 1995. *Of Bicycles, Bakelites, and Bulbs: Toward a Theory of Sociotechnical Change*. Cambridge, MA: MIT Press.

Blaise, Clark. 2000. *Time Lord: Sir Sanford Fleming and the Creation of Standard Time*. New York: Vintage.

Blumberg, S. J., and J. V. Luke. 2007. *Wireless Substitution: Early Release of Estimates from the National Health Interview Survey, January–June 2007*. Washington, DC: National Center for Health Statistics.

Bookchin, M. 1987. Social ecology versus deep ecology: A challenge for the ecology movement. *Green Perspectives: Newsletter of the Green Program Project* (1987): 4–5.

Boorstin, Daniel J. 1965. *The Americans: The National Experience*. New York: Vintage.

Borduin, C. M., C. M. Schaeffer, and N. Heiblum. 2009. A randomized clinical trial of multisystemic therapy with juvenile sexual offenders: Effects on youth social ecology and criminal activity. *Journal of Consulting and Clinical Psychology* 77 (1):26.

Boulding, K. E. 1955. An application of population analysis to the automobile population of the United States. *Kyklos* 8 (2):109–122.

Boulding, K. E. 1978. *Ecodynamics: A New Theory of Social Evolution*. London: Sage.

Boulding, K. E. 1980. *Beasts, Ballads, and Bouldingisms: A Collection of Writings by Kenneth E. Boulding*, ed. R. P. Beilock. New Brunswick, NJ: Transaction.

Boulding, K. E. 1991. What is evolutionary economics? *Journal of Evolutionary Economics* 1 (1):9–17.

Boyd, Dayna, and Nicole Ellison. 2007. Social network sites: Definition, history, and scholarship. *Journal of Computer-Mediated Communication* 13 (1):210–230.

Briscoe, Bob, Andrew Odlyzko, and Benjamin Tilly. 2006. Metcalfe's law is wrong—Communications networks increase in value as they add members—but by how much? *IEEE Spectrum* 43 (7):34–39.

Bronowski, J. 1973. *The Ascent of Man*. Boston: Little, Brown.

Brown, Katie, Scott W. Campbell, and Rich Ling. 2011. Mobile phones bridging the digital divide for teens in the US? *Future Internet 2011* (special issue, *Social Transformations from the Mobile Internet*) 3 (2):144–158.

Brundvand, Jan Harold. 1981. *The Vanishing Hitchiker: American Urban Legends and Their Meaning*. New York: Norton.

Bryson, B. 1994. *Made in America*. London: Black Swan.

Calhoun, C. 1992. The infrastructure of modernity: Indirect social relationships, information technology, and social integration. In *Social Change and Modernity*, ed. H. Haferkamp and N. Smelser. Berkeley: University of California Press.

Calhoun, Craig. 1998. Community without propinquity revisited: Communications technology and the transformation of the urban public sphere. *Sociological Inquiry* 68 (3):373–397.

Campbell, Scott, and T. C. Russo. 2003. The social construction of mobile telephony: An application of the social influence model to perceptions and uses of mobile phones within personal communication networks. *Communication Monographs* 70 (4):317–334.

Campbell, Scott, and Nojin Kwak. 2010. Mobile communication and social capital: An analysis of geographically differentiated usage patterns. *New Media & Society* 12 (3):435–451.

Carey, J. W. 1988. *Communication as Culture: Essays on Media and Society*. London: Routledge.

Carosso, V. P. 1949. The Waltham Watch Company: A case history. *Bulletin of the Business Historical Society* 23 (4):165–187.

Castells, Manuel. 2009. *Communication Power*. Oxford: Oxford University Press.

Castells, Manuel, Mereia Fernandez-Ardevol, Jack Qiu, and Araba Sey. 2007. *Mobile Communication and Society: A Global Perspective*. Cambridge, MA: MIT Press.

Chang, Stephanie E., and Nobuoto Nojima. 2001. Measuring post-disaster transportation system performance: The 1995 Kobe earthquake in comparative perspective. *Transportation Research, Part A: Policy and Practice* 35 (6):475–494.

Chayko, Mary. 2008. *Portable Communities: The Social Dynamics of Online and Mobile Connectedness*. Albany: SUNY Press.

Chen, C., and C. E. McKnight. 2007. Does the built environment make a difference? Additional evidence from the daily activity and travel behavior of homemakers living in New York City and suburbs. *Journal of Transport Geography* 15 (5):380–395.

Chen, Y.-F. 2007. *The Mobile Phone and Socialization: The Consequences of Mobile Phone Use in Transitions from Family to School Life of U.S. College Students*. New Brunswick, NJ: Rutgers University Press.

Cherry, C. 1977. The telephone system: Creator of mobility and social change. In *The Social Impact of the Telephone*, ed. I. de Sola Pool. Cambridge, MA: MIT Press.

Chesley, N. 2005. Blurring boundaries? Linking technology use, spillover, individual distress, and family satisfaction. *Journal of Marriage and the Family* 67:1237–1248.

Cohen, A., D. Lemish, and A. M. Schejter. 2007. *The Wonder Phone in the Land of Miracles: Mobile Telephony in Israel*. Cresskill, NJ: Hampton Press.

Cohen, A. A., and D. Lemish. 2005. When bombs go off the mobiles ring: The aftermath of terrorist attacks. In *A Sense of Place: The Global and the Local in Mobile Communication*, ed. K. Nyiri, 117–128. Vienna: Passagen Verlag.

Collins, Randall. 2004. *Interaction Ritual Chains*. Princeton: Princeton University Press.

Consalvo, Mia, and Charles Ess, eds. 2011. *The Handbook of Internet Studies*. London: Wiley.

Cowan, Ruth Schwartz. 1983. *More Work for Mother: The Ironies of Household Technology from the Open Hearth to the Microwave*. New York: Basic Books.

Crawford, M. 1994. The world in a shopping mall. In *Variations on a Theme Park: The End of Public Space*, ed. M. Sorkin, 3–30. New York: Hill & Wang.

Cumiskey, K. M. 2005. "Surprisingly, nobody tried to caution her": Perceptions of intentionality and the role of social responsibility in the public use of mobile phones. In *Mobile Communications: Re-Negotiation of the Social Sphere*, ed. R. Ling and P. Pedersen. Surrey: Springer-Verlag.

Davis, F. 1985. Clothing and fashion as communication. In *The Psychology of Fashion*, ed. M. R. Solomon, 15–27. Lexington: D. C. Heath.

de Gournay, C. 2002. Pretense of intimacy in France. In *Perpetual Contact: Mobile Communication, Private Talk, Pubic Performance*, ed. J. E. Katz and M. Aakhus. Cambridge: University of Cambridge Press.

DeGraff, Harmon O. 1926. The teaching of urban sociology. *Social Forces* 5 (1):248–254.

Denkel, A., and C. Ø. Madsen. 2010. Measuring the use of mobile phones in Denmark and the EU. Paper presented at Mobile Communication: Moving Forward. Copenhagen, Denmark, March 12, 2010.

Dilhac, J.-M. N.d. The telegraph of Claude Chappe—An optical telecommunication network for the 18th century. http://www.ieeeghn.org/wiki/images/1/17/Dilhac.pdf.

Dohrn-van Rossum, G. 1992. *History of the Hour: Clocks and Modern Temporal Orders*. Chicago: University of Chicago Press.

Donner, J. 2005. The rules of beeping: Exchanging messages using missed calls on mobile phones in Sub-Saharan Africa. Paper presented at the International Communications Association, New York.

Donner, J. 2007. The rules of beeping: Exchanging messages using missed calls on mobile phones in Sub-Saharan Africa. *Journal of Computer-Mediated Communication* 13 (1):1–22.

Donner, J. 2008. Research approaches to mobile use in the developing world: A review of the literature. *Information Society* 24 (3):140–159.

Donner, J., and S. Gitau. 2009. New paths: Exploring mobile-centric internet use in Cape Town. Paper presented at "Mobile 2.0: Beyond Voice?" Pre-conference workshop at the International Communication Association (ICA) Conference, Chicago, May 21, 2009.

Duncan, H. D. 1970. *Communication and the Social Order*. Oxford: Oxford University Press.

Durkheim, E. 1938. *The Rules of the Sociological Method*. New York: Free Press.

Durkheim, E. 1951. *Suicide*. New York: Free Press.

Durkheim, E. 1974. *Sociology and Philosophy*. New York: Free Press.

Durkheim, E. 1995. *The Elementary Forms of Religious Life. Translated by K.E. Fields*. Glencoe, IL: Free Press.

Dutton, W. 2003. The social dynamics of wireless on September 11: Reconfiguring access. In *Crisis Communication*, ed. A. M. Noll, 69–82. Lanham, MD: Rowan & Littlefield.

Dyckman, J. W. 1973. Transportation in the cities. In *Cities: Their Origin, Growth, and Human Impact*, ed. K. Davis, 195–206. San Francisco: Freeman.

Eckermann, E. 2001. *World History of the Automobile*. Washington, DC: SAE.

Ehrenreich, Barbara. 2008. *Nickel and Dimed: On (Not) Getting by in America*. New York: Holt Paperbacks.

Eisenstein, E. 1979. *The Printing Press as an Agent of Change: Communications and Cultural Transformations in Early-Modern Europe*, vols. 1 and 2. Cambridge: Cambridge University Press.

Elias, N. 1992. *Time: An Essay*. Oxford: Blackwell.

Elliott, A., and J. Urry. 2010. *Mobile Lives*. London: Routledge.

Ennett, S. T., V. A. Foshee, K. E. Bauman, A. Hussong, L. Cai, H. L. M. N. Reyes, R. Faris, J. Hipp, and R. DuRant. 2008. The social ecology of adolescent alcohol misuse. *Child Development* 79 (6):1777–1791.

Eriksen, T. H. 2001. *Tyranny of the Moment: Fast and Slow Time in the Information Age*. London: Pluto Press.

Eriksson, M. 2011. Trust in public safety answering points: A Swedish national survey in the late modern network society. *Journal of Homeland Security and Emergency Management* 8 (1):1–15.

Farley, T. 2005. Mobile telephone history. *Telektronikk* 3–4: 22–34.

Farley, T. N.d. Privateline.com: Telephone history. http://www.privateline.com/TelephoneHistory3A/numbers.html.

Fine, G. A. 1973. The Kentucky Fried rat: Legends and modern society. *Folklore Institute Journal* 17 (2/3):222–243.

Fine, G. A. 1979. Cokelore and Coke law: Urban belief tales and the problem of multiple origins. *Journal of American Folklore* 92 (366):477–482.

Fischer, C. 1976. *The Urban Experience*. San Diego: Harcourt Brace Jovanovich.

Fischer, C. 1982. *To Dwell among Friends: Personal Networks in Town and City*. Chicago: University of Chicago.

Fischer, C. 1992. *America Calling: A Social History of the Telephone to 1940*. Berkeley, CA: University of California Press.

Fischer, C. 2011. Home owning dreams. http://madeinamericathebook.wordpress.com/2011/05/18/home-owning-dreams/#more-1430.

Flichy, P. 2007. *The Internet Imaginaire*. Trans. Liz Carey-Libbrecht. Cambridge, MA: MIT Press.

Flink, J. J. 2001. *The Automobile Age*. Cambridge, MA: MIT Press.

Fortunati, L. 2005. Mobile telephone and the presentation of self. In *Mobile Communications: Re-Negotiation of the Social Sphere*, ed. R. Ling and P. Pedersen, 203–218. London: Springer.

Fortunati, L. 1998. The ambiguous image of the mobile phone. In *Communications on the Move: The Experience of Mobile Telephony in the 1990s*, ed. L. Fortunati. Farsta: Telia.

Fortunati, L. 2003. Mobile phone and the presentation of self. Paper presented at Front Stage/Back Stage: Mobile Communication and the Renegotiation of the Social Sphere, Grimstad, Norway, June 22–24, 2003.

Foster, M. 1983. The automobile and the city. In *The Automobile and American Culture*, ed. David L. Lewis and Lawrence Goldstein, 24–36. Ann Arbor: University of Michigan Press.

Friedman, A. J. 1984. The clockwork universe. *Technology and Culture* 25 (2):280–286.

Frye, H. 1983. The automobile and American fashion, 1900–1930. In *The Automobile and American Culture*, ed. D. L. Lewis and L. Goldstein, 48–58. Ann Arbor: University of Michigan Press.

Gamson, W. 1975. *The Strategy of Social Protest*. Chicago: Dorsey.

Ganem, Jacques, and Culley Carson. 1999 Frere Jacques Beaulieu: From rogue lithotomist to nursery rhyme character. *Journal of Uruology* 161 (4):1067–1069.

Gans, Herbert J. 1967. *The Levittowners: Ways of Life and Politics in a New Suburban Community*. New York: Vintage.

Geertz, Clifford. 1972. Linguistic etiquette. In *Readings in the Sociology of Language*, ed. Joshua A. Fishman, 282–295. The Hague: Mouton.

Geirbo, H. C., P. Helmersen, and K. Engø-Monsen. 2007. Missed call: Messaging for the masses. A Study of Missed Call Signaling Behavior in Dhaka. (Internal Telenor R&I Publication.) Fornebu: Telenor R&I.

Giddens, A. 1991. *Modernity and Self-Identity: Self and Society in the Late Modern Age*. Cambridge: Polity.

Giddens, A. 1986. *The Constitution of Society: Outline of the Theory of Structuration*. Berkeley: University of California Press.

Gilbert, M. 1989. *On Social Facts*. London: Routledge.

Gittu, J., I. Jørgensen, and S. Nørve. 1985. *Bovaner: En Undersøkelse Av 30 Blokkleiligheter I Oslo*. Oslo: Byggforsk.

Gleick, J. 2011. *The Information: A History, a Theory, a Flood*. New York: Pantheon.

Glennie, P., and N. Thrift. 2009. *Shaping the Day: A History of Timekeeping in England and Wales 1300–1800*. Oxford: Oxford University Press.

Goffman, E. 1963. *Behavior in Public Places: Notes on the Social Organization of Gatherings*. New York: Free Press.

Goffman, E. 1967. *Interaction Ritual: Essays on Face-to-Face Behavior*. New York: Pantheon.

Goggin, G. 2009. Adapting the mobile phone: The iPhone and its consumption. *Continuum: Journal of Media and Cultural Studies* 23 (2):231–244.

Goodwin, D. K. 1994. *No Ordinary Time*. New York: Simon & Schuster.

Gordon, E., and A. de Souza e Silva. 2011. *Net Locality: Why Location Matters in a Networked World*. Oxford: Wiley-Blackwell.

Gottman, J. 1977. Megalopolis and antipolis: The telephone and the structure of the city. In *The Social Impact of the Telephone*, ed. I. de Sola Pool. Cambridge, MA: MIT Press.

Grazioli, Stefano, and Ronald C. Carrell. 2002. Exploding phones and dangerous bananas: Perceived precision and believability of deceptive messages found on the Internet. Paper presented at Eighth Americas Conference on Information Systems, Dallas, Texas, August 9–11, 2002.

Green, Nicola. 2002. On the move: Technology, mobility, and the mediation of social time and space. *Information Society* 18 (4):281–292.

Greenhut, Steven. 2003. You can have my new SUV when you peel my cold, dead hands off its leather steering wheel. http://www.lewrockwell.com/orig3/greenhut2.html.

Guillory, D. L. 1983. Bel Air: The automobile as art object. In *The Automobile and American Culture*, ed. D. L. Lewis and L. Goldstein, 280–289. Ann Arbor: University of Michigan Press.

Gullestad, M. 1992. *The Art of Social Relations: Essays on Culture, Social Action, and Everyday Life in Modern Norway*. Oslo: Universitetetsforlaget.

Gunthorpe, W., and K. Lyons. 2004. A predictive model of chronic time pressure in the Australian population: Implications for leisure research. *Leisure Sciences* 26 (2):201–213.

Habuchi, I. 2005. Accelerating reflexivity. In *Personal, Portable, Pedestrian: Mobile Phones in Japanese Life*, ed. M. Ito, D. Okabe, and M. Matsuda, 165–182. Cambridge, MA: MIT Press.

Haddon, L. 2003. Domestication and mobile telephony. In *Machines That Become Us*, ed. James E. Katz, 43–56. New Brunswick, NJ: Transaction.

Haddon, L. 2004. *Information and Communication Technologies in Everyday Life*. Oxford: Berg.

Haddon, L. 2006. The contribution of domestication research to in-home computing and media consumption. *Information Society* 22 (4):195–203.

Hall, J., and N. Baym. 2011. Calling and texting (too much): Mobile maintenance expectations, (over)dependence, entrapment, and friendship satisfaction. *New Media & Society* 13 (6).

Hall, P. 1996. *Cities of Tomorrow: An Intellectual History of Urban Planning and Design in the 20th Century*. Oxford: Blackwell.

Hampton, K., L. Sessions, E. J. Her, and L. Rainie. 2009. *Social Isolation and New Technology*. Washington, DC: Pew Internet and American Life Project.

Hampton, K., and B. Wellman. 2003. Neighboring in Netville: How Internet supports community and social capital in a wired suburb. *City & Community* 2 (4):277–311.

Hargittai, E., J. Gallo, and M. Kane. 2008. Cross-ideological discussions among conservative and liberal bloggers. *Public Choice* 134 (1–2):67–86.

Hayden, Dolores. 2003. *Building Suburbia: Greenfields and Urban Growth 1820–2000*. New York: Vintage.

Helles, Rasmus. 2009. *Personlige Medier I Hverdagslivet*. Københavns Universitet.

Hey, Kenneth. 1983. Cars and films in American culture. In *The Automobile and American Culture*, ed. D. L. Lewis and L. Goldstein, 193–205. Ann Arbor: University of Michigan Press.

Hillebrand, Friedhelm, Finn Trosby, Kevin Holley, and Ian Harris. 2010. *Short Message Service: The Creation of Personal Global Text Messaging*. Chichester: Wiley.

Hjorthol, R. J. 2000. Same city—different options: An analysis of the work trips of married couples in the metropolitan area of Oslo. *Journal of Transport Geography* 8:213–220.

Hjorthol, R. J., M. H. Jakobsen, R. Ling, and S. Nordbakke. 2007. Det Mobile Hverdagsliv: Kommunikasjon Og Koordinering I Moderne Barnefamilier. In *Personlige Medier: Livet Mellom Skjermene*, ed. M. Lüders, L. Prøitz, and T. Rasmussen. Oslo: Gyldendal Akademisk.

Hollenbaugh, Erin E., and Amber L. Ferris. 2010. "i love you, man": Drunk Dialing motives and their impact on social cohesion. In *Mobile Communication: Bringing Us Together or Tearing Us Apart? The Mobile Communication Research Series*, vol. 2, ed. R. Ling and S. Campbell. New Brunswick: Transaction.

Horst, Heather A., and Daniel Miller. 2005. From kinship to link-up: Cell phones and social networking in Jamaica. *Current Anthropology* 46 (5):755–778.

Howe, D. W. 2007. *What Hath God Wrought: The Transformation of America, 1815–1848*. Oxford: Oxford University Press.

Hulme, Michael, and Anna Truch. 2006. The role of interspace in sustaining identity. *Knowledge, Technology, & Policy* 19 (1):45–53.

Humphreys, L. 2007. *Mobile Sociality and Spatial Practice: A Qualitative Field Study of New Social Networking Technologies*. Philadelphia: University of Pennsylvania Press.

Hunsinger, J., L. Klastrup, and M. Allen, eds. 2010. *International Handbook of Internet Research*. Heidelberg: Springer.

Igarashi, T., J. Takai, and T. Youhida. 2005. A longitudinal study of social network development via mobile phone text messages focusing on gender differences. *Journal of Social and Personal Relationships* 22 (4):691–713.

Inglis, D. 2009. Cosmopolitan sociology and the classical canon: Ferdinand Tönnies and the emergence of global Gesellschaft. *British Journal of Sociology* 60 (4):813–832.

Interrante, J. 1983. The road to autopia: The automobile and the spatial transformation of American culture. In *The Automobile and American Culture*, ed. D. L. Lewis and L. Goldstein, 89–104. Ann Arbor: University of Michigan Press.

Ishii, K. 2006. Implications of mobility: The uses of personal communication media in everyday life. *Journal of Communication* 56:346–365.

Ito, M. 2005. Introduction: Personal, portable, pedestrian. In *Personal, Portable, Pedestrian: Mobile Phones in Japanese Life*, ed. M. Ito, D. Okabe, and M. Matsuda, 1–16. Cambridge, MA: MIT Press.

Ito, M., D. Okabe, and M. Matsuda, eds. 2005. *Personal, Portable, Pedestrian: Lessons from Japanese Mobile Use*. Cambridge, MA: MIT Press.

Ito, M., and D. Okabe. 2005. Intimate connections: Contextualizing Japanese youth and mobile messaging. In *The Inside Text: Social, Cultural and Design Perspectives on SMS*, ed. R. Harper, L. Palen, and A. Taylor, 127–145. Dordrecht: Springer.

Ito, M., S. Baumer, M. Bittanti, d. boyd, R. Cody, B. Herr-Stephenson, H. A. Horst, P. G. Lange, D. Mahendran, K. Z. Martínez, C. J. P. D. Perkel, L. Robinson, C. Sims, and L. Tripp. 2010. *Hanging Out, Messing Around, and Geeking Out: Kids Living and Learning with New Media*. Cambridge, MA: MIT Press.

ITU (International Telecommunications Union). 2009. Market information and statistics. http://www.itu.int/ITU-D/ict/statistics/.

ITU (International Telecommunications Union). 2010. Global ICT development, 2000–2010. http://www.itu.int/ITU-D/ict/statistics/material/graphs/2010/Global_ICT_Dev_00-10.jpg.

ITU-D (International Telecommunications Union). 2010. The world in 2010: The rise of 3g. http://www.itu.int/ITU-D/ict/material/FactsFigures2010.pdf.

Jackson, K. T. 1985. *Crabgrass Frontier: The Suburbanization of the United States*. New York: Oxford University Press.

Jensen, R. 2007. The digital provide: information (technology), market performance, and welfare in the South Indian fisheries sector. *Quarterly Journal of Economics* 122 (3):879–924.

Johannesen, S. 1981. *Sammendrag Av Markedsundersøkelser Gjennomført for Televerket I Tiden 1966–1981*. Kjeller: Televerkets Forskninginstitutt.

Kahlert, H., R. Mühe, and G. L. Brunner. 1986. *Wristwatches: History of a Century's Development*. West Chester, PA: Schiffer.

Kalba, K. 2008. The adoption and diffusion of mobile phones—Nearing the halfway mark. *International Journal of Communication* 2:631–661.

Kalba, K. 2007. The adoption of mobile phones in emerging markets: Global diffusion and the rural challenge. Paper presented at the 6th Annual Global Mobility Roundtable, Center for Telecom Management, Marshall School of Business. Los Angeles: University of Southern California, June 1–2, 2007.

Kalil, T. 2009. Harnessing the mobile revolution. *Innovations: Technology, Governance, Globalization* 4 (1):9–23.

Katz, J. E. 2008. Mainstream mobiles in daily life: Perspectives and prospects. In *Handbook of Mobile Communication Studies*, ed. J. Katz, 433–445. Cambridge, MA: MIT Press.

Katz, J. E., and M. Aakhus, eds. 2002. *Perpetual Contact: Mobile Communication, Private Talk, Public Performance*. Cambridge: Cambridge University Press.

Katz, J. E., and R. Rice. 2002a. *Social Consequences of Internet Use*. Cambridge, MA: MIT Press.

Katz, J. E., and R. Rice. 2002b. The telephone as an instrument of faith, hope, terror, and redemption: America, 9-11. *Promethius* 20 (3):247–253.

Kihlstedt, F. T. 1983. The automobile and the transformation of the American house, 1910–1935. In *The Automobile and American Culture*, ed. D. L. Lewis and L. Goldstein, 160–175. Ann Arbor: University of Michigan Press.

Kim, H., G. J. Kim, H. W. Park, A. Sogang, and Y. Nam. 2006. The configurations of social relationships in communication channels: F2f, email, messenger, mobile phone, and SMS. Paper presented at the ICA Pre-conference on Mobile Communication, Erfert/Dresden, Germany, June 14–24.

Kovanen, L., J. Saramäki, and K. Kaski. 2011. Reciprocity of mobile phone calls. *Dynamics of Socio-Economic Systems* 2 (2):138–151.

Kristeva, J. 1981. Woman's time. *Signs* 7:13–35.

Kwan, M. P. 2006. New information technologies: Human behavior in space-time and the urban economy. In *Information Technology at Home*, ed. R. Kraut. Oxford: Oxford University Press.

Landes, D. S. 1983. *Revolution in Time: Clocks and the Making of the Modern World*. Cambridge, MA: Harvard University Press.

Lasén, A. 2011. "Mobiles are not that personal": The unexpected consequences of the accountability, accessibility, and transparency afforded by mobile telephony. In *Mobile Communication: Bringing Us Together or Tearing Us Apart? The Mobile Communication Research Series*, volume 2, ed. R. Ling and S. Campbell. New Brunswick: Transaction.

Lash, S. 2002. Individualization in a non-linear mode. In *Individualization*, ed. U. Bech and E. Beck-Gernsheim, vii–xiii. London: Sage.

Leena, K., L. Tomi, and R. Arja. 2005. Intensity of mobile phone use and health compromising behaviours—How is information and communication technology connected to health-related lifestyle in adolescence? *Journal of Adolescence* 28 (1):35–47.

Lenhart, A., M. Madden, A. R. Macgill, and A. Smith. 2007. Teens and social media. Washington, DC: Pew Internet and American Life Project.

Lenhart, A., R. Ling, S. Campbell, and K. Purcell. 2010. Teens and mobile phones. Washington, DC: Pew Internet and American Life Project.

Leonard, D., and S. Sensiper. 1998. The role of tacit knowledge in group innovation. *California Management Review* 40 (3):112–132.

Levine, R. V. 1998. *A Geography of Time: The Temporal Misadventures of a Social Psychologist, or How Every Culture Keeps Time Just a Little Bit Differently*. New York: Basic.

Levinson, M. 2011. *The Great A&P and the Struggle for Small Business in America*. New York: Hill & Wang/Farrar, Straus & Giroux.

Lewis, D. L. 1983. Sex and the automobile: From rumble seats to rockin' vans. In *The Automobile and American Culture*, ed. D. L. Lewis and L. Goldstein. Ann Arbor: University of Michigan Press.

Licoppe, C. 2007. Co-proximity events: Weaving mobility and technology into social encounters. Paper presented at Towards a Philosophy of Telecommunications Convergence, Budapest, September 27–29.

Licoppe, Christian. 2004. Connected presence: The emergence of a new repertoire for managing social relationships in a changing communications technoscape. *Environment and Planning: Society and Space* 22:135–156.

Lienert, R. M. 1983. Moving backward. In *The Automobile and American Culture*, ed. D. L. Lewis and L. Goldstein, 160–175. Ann Arbor: University of Michigan Press.

Lin, J. C. C., and C. L. Hsu. 2009. A multi-faceted analysis of factors affecting the adoption of multimedia messaging service (MMS). *International Journal of Technology and Human Interaction* 5 (4):18–36.

Ling, R. 1997. "One can talk about common manners!": The use of mobile telephones in inappropriate situations. In *Themes in Mobile Telephony Final Report of the Cost 248 Home and Work Group*, ed. L. Haddon, 73–96. Stockholm: Telia.

Ling, R. 2002. The social juxtaposition of mobile telephone conversations and public spaces. Paper presented at The Social Consequences of Mobile Telephones, Chunchon, Korea, July 13–15, 2002.

Ling, R. 2004. *The Mobile Connection: The Cell Phone's Impact on Society*. San Francisco: Morgan Kaufmann.

Ling, R. 2005. The socio-linguistics of SMS: An analysis of SMS use by a random sample of Norwegians. In *Mobile Communications: Renegotiation of the Social Sphere*, ed. R. Ling and P. Pedersen, 335–349. London: Springer.

Ling, R. 2008a. Exclusion or self-isolation? Texting and the elderly users. *Information Society* 24 (5):334–341.

Ling, R. 2008b. *New Tech, New Ties: How Mobile Communication Is Reshaping Social Cohesion*. Cambridge, MA: MIT Press.

Ling, R. 2010. Texting as a life phase medium. *Journal of Computer-Mediated Communication* 15 (2):277–292.

Ling, R., and N. Baron. 2007. The mechanics of text messaging and instant messaging among American college students. *Journal of Language and Social Psychology* 26 (3):291–298.

Ling, R., T. Bertel, and P. Sundsøy. 2012. The socio-demographics of texting: An analysis of traffic data. *New Media & Society*.

Ling, R., J. Bjelland, P. Sundsøy, and S. Campbell. 2011. Small circles: Mobile telephony and the cultivation of the private sphere. Paper presented at International Communication Association, Boston, Massachusetts, May 22–26, 2011.

Ling, R., and J. Donner. 2009. *Mobile Communication*. London: Polity.

Ling, R., K. Engø-Monsen, P. Sundsøy, J. Bjelland, and G. Canright. Forthcoming. Small and even smaller circles: The size of mobile phone based core social networks in Scandinavia and South Asia. *Journal of Intercultural Research.*

Ling, R., T. Julsrud, and E. Krogh. 1998. The Goretex principle: The Hytte and mobile telephones in Norway. In *Communications on the Move: The Experience of Mobile Telephony in the 1990s* (Cost248 Report), ed. L. Haddon. Farsta: Telia.

Ling, R., and R. N. McEwen. 2010. Mobile communication and ethics: Implications of everyday actions on social order. *Etikk i praksis: Nordic Journal of Applied Ethics* 4 (2):11–26.

Ling, R., and G. B. Stald. 2010. Mobile communities: Are we talking about a village, a clan, or a small group? *American Behavioral Scientist* 53 (4).

Ling, R., and P. Sundsøy. 2009. The iPhone and mobile access to the Internet. In *Mobile Media and the Change of Everyday Life*, ed. J. Höflich, C. Dietmar, G. Kircher, and I. Schlote. Berlin: Peter Lang.

Ling, R., and P. Sundsøy. Forthcoming. Mobile communication as a mundane medium. In *The Philosophy of Mobile Communication*, ed. K. Nyíri. Vienna: Passagen Verlag.

Ling, R., and D. Svanæs. 2011. Browsers vs. apps: The role of apps in the mobile internet. Paper presented at Internet and Society: Challenges, Transformation, and Development. University of Beijing, May 2011.

Ling, R., and B. Yttri. 2002. Hyper-coordination via mobile phones in Norway. In *Perpetual Contact: Mobile Communication, Private Talk, Public Performance*, ed. J. E. Katz and M. Aakhus, 139–169. Cambridge: Cambridge University Press.

Ling, R., B. Yttri, B. Anderson, and D. DeDuchia. 2003. Mobile communication and social capital in Europe. In *Mobile Democracy: Essays on Society, Self and Politics*, ed. K. Nyri. Vienna: Passagen Verlag.

Livingstone, Sonia, Leslie Haddon, Anke Görzig, and Kjartan Ólafsson. 2011. *Risks and Safety on the Internet*. London: London School of Economics.

Livingstone, S. 2008. On the mediation of everything: ICA Presidential Address 2008. *Journal of Communication* 59 (1):1–18.

Lou, H., W. Luo, and D. Strong. 2000. Perceived critical mass effect on groupware acceptance. *European Journal of Information Systems* 9 (2):91–103.

Loudon, I. 2002. *Western Medicine: An Illustrated History*. Oxford: Oxford University Press.

Lynd, R. S., and H. M. Lynd. 1929. *Middletown: A Study in Modern American Culture*. New York: Harcourt Brace.

MacKenzie, A. 2002. *Transductions: Bodies and Machines at Speed*. London: Continuum International Publishing Group.

MacKenzie, D., and J. Wajcman, eds. 1985. *The Social Shaping of Technology: How the Refrigerator Got Its Hum*. Milton Keynes: Open University Press.

MacKenzie, S., H. Kober, D. Smith, T. Jones, and E. Skepner. 2001. Letterwise: Prefix-based disambiguation for mobile text input. In *Proceedings of UIST 2001*, 111–120. New York: ACM.

Maitland, D. 1984. *The Missing Link: Report of the Independent Commission for World-Wide Telecommunications Development*. Geneva: ITU.

Manning, P. K. 1996. Information technology in the police context: The "sailor" phone. *Information Systems Research* 7 (1):52–62.

Markus, L. M. 1987. Toward a "critical mass" theory of interactive media: Universal access, interdependence, and diffusion. *Communication Research* 14 (5):491–511.

Martin, M. 1991. *Hello, Central? Gender, Technology and Culture in the Formation of Telephone Systems*. Montreal: McGill-Queens.

Marvin, C. 1988. *When Old Technologies Were New: Thinking about Electric Communication in the Late Ninteenth Century*. New York: Oxford University Press.

Matsuda, M. 2005. Mobile communication and selective sociality. In *Personal, Portable, Pedestrian: Mobile Phones in Japanese Life*, ed. M. Ito, D. Okabe, and M. Matsuda, 123–142. Cambridge, MA: MIT Press.

McEwen, R. N. 2009. A world more intimate: Exploring the role of mobile phones in maintaining and extending social networks. Docotoral dissertation. University of Toronto, Faculty of Information.

McPherson, M., L. Smith-Lovin, and M. E. Brashears. 2006. Social isolation in America: Changes in core discussion networks over two decades. *American Sociological Review* 71:353–375.

Millard-Ball, A., and L. Schipper. 2010. Are we reaching a plateau or a "peak" travel? Trends in passenger transport in six industralized countries. Paper presented at Transportation Research Board Annual Meeting, Washington, DC, January 10–14, 2010.

Miyata, K. 2006. Longitudinal effects of mobile internet use on social network in Japan. Paper presented at the The International Communications Association Conference, Dresden, June 14–24, 2006.

Mok, D., and B. Wellman. 2007. Did distance matter before the Internet? Interpersonal contact and support in the 1970s. *Social Networks* 29 (3):430–461.

Moline, N. C. 1971. Mobility and the Small Town, 1900–1930: Transportation Change in Oregon, Illinois. University of Chicago, Department of Geography. Research paper no. 12.

Money, H. A., and R. J. Hobbs. 2000. *Invasive Species in a Changing World*. Washington, DC.: Island Press.

Monk, A., J. Carroll, S. Parker, and M. Blythe. 2004. Why are mobile phones annoying? *Behaviour & Information Technology* 23 (1):33–41.

Mumford, L. 1963. *Technics and Civilization*. San Diego: Harvest.

Nesvold, B. 2006. "Fortell Hvem Du Er!": Mobiltelefonen Som Ledd I Ungdoms Identitetskonstruksjon. Master's thesis, Norges Teknisk-naturvitenskaplig Universitetet.

Nielsen, J. 2006. Participation inequality: Encouraging more users to contribute. http://www.useit.com/alertbox/participation_inequality.html.

Nordal, K. 2000. Takt Og Tone Med Mobiltelefon: Et Kvalitativt Studie Om Folks Brug Og Opfattelrer Af Mobiltelefoner. Master's thesis, Universitetet i Oslo.

Nordsveen, A. M. 2006. *Mobilen: Mobiltelefoner I Norge 1966–2006*. Oslo: Norsk Telemuseum.

Novak, Jakub, and Ludek Sykora. 2007. A city in motion: Time-space activity and mobility patterns of suburban inhabitants and the structuration of the spatial organization of the Prague metropolitan area. *Geografiska Annaler, Series B: Human Geography* 89 (2):147–168.

O'Connor, K., and C. Maher. 1982. Change in the spatial structure of a metropolitan region: Work-residence relationships in Melbourne. In *Internal Structure of the City: Readings from Urban Form, Growth and Policy*, ed. L. S. Bourne, 406–421. New York: Oxford University Press.

Palen, L. 1998. *Calendars on the New Frontier: Challenges of Groupware Technology*. Irvine: University of California Press.

Paragas, F. 2003. Dramatextism: Mobile telephony and people power in the Philippines. In *Essays on Society, Self, and Politics*, ed. K. Nyíri. Vienna: Passagen Verlag.

Park, W. K. 2005. Mobile telephone addiction. In *Mobile Communications: Renegotiation of the Social Sphere*, ed. R. Ling and P. Pedersen, 259–272. London: Springer.

Pierce, Albert. 1960. Durkheim and functionalism. In *Emile Durkheim*, ed. Kurt Wolff, 154–156. Columbus: Ohio State University Press.

Putnam, R. 2000. *Bowling Alone: The Collapse and Revival of American Community*. New York: Touchstone.

Qiu, J. L., and C. Wallis. 2009. Shanzhaiji and the transformation of local mediascape in Shenzhen. In *Provincial China: Provincial and Local Media*. Hefei, Anhui Province: University of Technology Sydney.

Rafael, V. L. 2003. The cell phone and the crowd: Messianic politics in the contemporary Philippines. *Public Culture* 15 (3):399–425.

Rahman, T. 2007. "Real markets" in rural Bangladesh: Institutions, market interactions, and the reproduction of inequality. In *IPPG Briefing Paper No. Eight, in Conjunction with CUTS International*. Manchester: IPPG, University of Manchester.

Rakow, L. F. 1988. Women and the telephone: The gendering of a communications technology. In *Technology and Women's Voices: Keeping in Touch*, ed. C. Kramarae, 207–229. New York: Routledge.

Rakow, L. F., and V. Navarro. 1993. Remote mothering and the parallel shift: Women meet the cellular telephone. *Critical Studies in Mass Communication* 10:144–157.

Reid, D., and F. Reid. 2004. Insights into the social and psychological effects of SMS text messaging. http://www.160characters.org/documents/SocialEffectsOfTextMessaging.pdf.

Reis, H. T. 1980. Perceptions of time and punctuality in the United States and Brazil. *Journal of Personality and Social Psychology* 38 (4):541–550.

Reisner, Marc. 1986. *Cadillac Desert: The American West and Its Disappearing Water*. New York: Penguin.

Rheingold, Howard. 2002. *Smart Mobs*. Cambridge, MA: Perseus.

Rice, Ron. 1982. Communication networking in computer conferencing systems: A longitudinal study of group roles and system structure. In *Communication Yearbook*, ed. M. Burgoon, 925–944. Newbury Park, CA: Sage.

Richards, P. 2009. Dressed to kill: Clothing as technology of the body in the civil war in Sierra Leone. *Journal of Material Culture* 14 (4):495.

Riesman, D., R. Denney, and N. Glazer. [1950] 2001. *The Lonely Crowd: A Study in the Changing American Character*. New Haven: Yale University Press.

Robinson, L. M.. 1990. Regulating what we eat: Mary Engle Pennington and the food research laboratory. *Agricultural History* 64 (2):143–153.

Rogers, E. 1995. *Diffusion of Innovations*. New York: Free Press.

Rose, A. C. 1976. *Historic American Roads: From Frontier Trails to Superhighways*. New York: Crown Publishers.

Rosen, C. 2008. Everyone is talking. *Wall Street Journal*, April 21.

Rubin, L. 1985. *Just Friends: The Role of Friendship in Our Lives*. New York: Harper.

Sandvig, C., and H. Sawhney. 2004. Approaching yet another new communication technology. Paper presented at the Wireless Communication Workshop: Inspirations from Unusual Sources, Ann Arbor, Michigan, August 27–29, 2004.

Sanford, C. L. 1983. Women's place in American car culture. In *The Automobile and American Culture*, ed. D. L. Lewis and L. Goldstein, 137–152. Ann Arbor: University of Michigan Press.

Schipper, L., F. Unander, S. Murtishaw, and M. Ting. 2001. Indicators of energy use and carbon emissions: Explaining the energy economy link. *Annual Review of Energy and the Environment* 26:49–81.

Schroeder, R., and R. Ling. Forthcoming. Durkheim and Weber on the social implications of new information and communication technologies.

Schroeder, R. 2007. *Rethinking Science, Technology, and Social Change*. Stanford: Stanford University Press.

Setterberg, F. 1983. Crusing with Donny on the San Leandro Strip. In *The Automobile and American Culture*, ed. D. L. Lewis and L. Goldstein, 315–323. Ann Arbor: University of Michigan Press.

Shaw, J. 1994. Punctuality and the everyday ethics of time: Some evidence from the Mass Observation Archive. *Time & Society* 3 (1):79–97.

Sheller, M., and J. Urry. 2006. The new mobilities paradigm. *Environment and Planning* 38:207–226.

Sicart, M. 2009. *The Ethics of Computer Games*. Cambridge, MA: MIT Press.

Silk, G. 1983. The image of the automobile in American art. In *The Automobile and American Culture*, ed. D. L. Lewis and L. Goldstein, 206–221. Ann Arbor: University of Michigan Press.

Silverstone, R., and L. Haddon. 1996. Design and domestication of information and communication technologies: Technical change and everyday life. In *Communication by Design: The Politics of Information and Communication Technologies*, ed. R. Silverstone and R. Mansell, 44–74. Oxford: Oxford University Press.

Silverstone, R., E. Hirsch, and D. Morley. 1992. Information and communication technologies and moral economy of the household. In *Consuming Technologies: Media and Information in Domestic Spaces*, ed. R. Silverstone and E. Hirsch, 15–31. London: Routledge.

Simmel, G. [1903] 1971. The metropolis and mental life. In *Georg Simmel: On Individuality and Social Forms*, ed. D. N. Levine, 324–399. Chicago: University of Chicago Press.

Smith, D. H. 1973. Old cars and social productions among the Teton Lakota. Doctoral dissertation, Department of Sociology, University of Colorado.

Smoreda, Z., and F. Thomas. 2001. Social networks and residential ICT adoption and use. Paper presented at EURESCOM Summit 2001, 3G Technologies and Applications, Heidelberg, November 12–15.

Snider, B. L., and D. Sheals. 2003. Route 66 in Missouri. *National Park Service, Route 66 Preservation Program,* http://www.nps.gov/history/rt66/histsig/missouricontext.htm#_ftnref72.

Sobel, D. 1996. *Longitude.* London: Fourth Estate.

Song, C., Z. Qu, N. Blumm, and A.-L. Barabasi. 2010. Limits of predictability in human mobility. *Science* 327:1018–1021.

Sørensen, K. H., and P. Østby. 1995. *Flukten Fra Detroit: Bilens Integrasjon I Det Norske Samfunnet.* Trondheim: Senter for teknologi og samfunn.

Sprague, P. E. 1981. The origin of balloon framing. *Journal of the Society of Architectural Historians* 40(4):311–319.

Stald, G. 2007. Mobile monitoring: Aspects of risk and surveillance and questions of democratic perspectives in young people's uses of mobile phones. In *Young Citizens and New Media: Strategies of Learning for Democratic Engagement,* ed. P. Dahlgren. London: Routledge.

Standage, T. 1998. *The Victorian Internet.* London: Weidenfeld & Nicolson.

Star, S. L., and G. C. Bowker. 2006. How to infrastructure. *Handbook of New Media: Social Shaping and Social Consequences of ICTs* 230:151–162.

Stokols, D. 1992. Establishing and maintaining healthy environments: Toward a social ecology of health promotion. *American Psychologist* 47 (1):6.

Strayer, D. L., F. A. Drews, and D. J. Crouch. 2006. A comparison of the cell phone driver and the drunk driver. *Human Factors* 48 (2):381–392.

Sundsøy, P. R., J. Bjelland, G. Canright, K. Engø-Monsen, and R. Ling. 2010. Time evolution of products in social networks: An empirical study from a telecom market. Paper presented at the International Conference on Advances in Social Network Analysis and Mining, Odense, Denmak, August 9–11, 2010.

Sundsøy, P. R., J. Bjelland, G. Canright, K. Engø-Monsen, and R. Ling. In process. The mobilization of social networks in the wake of the 22 July bombing in Oslo, Norway.

Sutko, D. M., and A. de Souza e Silva. 2011. Location-aware mobile media and urban sociability. *New Media & Society* 13 (5):807–923.

Taylor, P., C. Funk, and A. Clark. 2006. Luxury or necessity? Things we can't live without: The list has grown in the past decade. http://pewsocialtrends.org/assets/pdf/Luxury.pdf.

Temin, P., and L. Galambos. 1987. *The Fall of the Bell System*. Cambridge: Cambridge University Press.

Thompson, E. P. 1967. Time, work discipline, and industrial capitalism. *Past & Present* 38:65–97.

Thorns, D. C. 1972. *Suburbia*. London: MacGibbon & Kee.

Toby, J. 1952. Some variables in role conflict analysis. *Social Forces* 30:323–327.

Tönnies, F. 1965. Gemeinschaft and Gesellschaft. In *Theories of Society*, ed. T. Parsons, E. Shils, K. D. Naegele, and J. R. Pitts, 191–201. New York: Free Press.

Tönnies, F. 1974. Gemeinschaft and Gesellschaft. In *The Sociology of Community: A Selection of Readings*, ed. Colin Bell and Howard Newby, 7–12. London: Routledge.

Tönnies, F. 2002. *Community and Society*. Newton Abbot, Devon: Courier Dover.

Tuchman, B. 1978. *A Distant Mirror: The Calamitous 14th Century*. New York: Ballentine.

Townsend, A. M. 2000. Life in the real time city—Mobile telephones and urban metabolism. *Journal of Urban Technology* 7:85–104.

Traugott, M., S.-H. Joo, R. Ling, and Y. Qian. 2006. On the move: The role of cellular communication in American life. Pohs Report on Mobile Communication. Ann Arbor, MI: University of Michigan, Department of Communication Studies.

Trosby, F. 2004. SMS, the strange duckling of GSM. *Telektronikk* 3:187–194.

Urry, J. 2007. *Mobilities*. Cambridge: Polity.

Urry, J. 2000. *Sociology beyond Societies: Mobilities for the Twenty-First Century*. London: Routledge.

Vaage, O. 2008. *Mediabruks Undersøkelse, 2007*. Oslo: Statistics Norway.

Vaage, O. 2011. *Mediabruks Undersøkelse, 2010*. Oslo: Statistics Norway.

Veach, S. R. 1981. *Children's Telephone Conversations*. University Microfilms International.

Wallis, C. 2011. Mobile phones without guarantees: The promises of technology and the contingencies of culture. *New Media & Society* 13 (3):471–485.

Wallis, C., J. Qiu, and R. Ling. 2012. Mobile handsets from the bottom up: Appropriation and innovation in the Global South. In *Media Studies Futures*, ed. K. Gates. London: Blackwell.

Walton, M., and J. Donner. 2009. Read-write-erase: Mobile-mediated publics in South Africa's 2009 elections. Paper presented at the International Conference on Mobile Communication and Social Policy, Rutgers University, New Brunswick, New Jersey, October 9–11, 2009.

Weber, M. 2002. *The Protestant Ethic and the Spirit of Capitalism*. London: Routledge.

Wei, Ran, and Ven-Hwei Lo. 2006. Staying connected while on the move: Cell phone use and social connectedness. *New Media & Society* 8 (1):53–72.

Weiner, E. 2007. Our cell phones, ourselves. http://www.npr.org/templates/story/story.php?storyId=17486953.

Wellman, B. 1999. The network community. In *Networks in the Global Village*, ed. B. Wellman, 1–48. Boulder: Westview.

Wellman, B. 2002. Little boxes, glocalization, and networked individualism. In *Digital Cities II: Computational and Sociological Approaches*, ed. M. Tanabe, P. van den Besselaar, and T. Ishida, 10–25. Berlin: Springer.

Wellman, B., and C. Haythornthwaite, eds. 2002. *The Internet in Everyday Life*. Oxford: Blackwell.

Wellman, B., B. Hogan, K. Berg, J. Boase, J. A. Carrasco, R. Côté, J. Kayhara, T. L. M. Kennedy, and P. Tran. 2005. Connected lives: The project. In *Networked Neighborhoods*, ed. P. Purcell, 161–216. Berlin: Springer.

Whitrow, G. J. 1988. *Time in History: Views of Time from Prehistory to the Present Day*. Oxford: Oxford University Press.

Williams, R. 1974. *Television: Technology and Cultural Form*. London: Fontana.

Wurtzel, A. H., and C. Turner. 1977. Latent functions of the telephone: What missing the extension means. In *The Social Impact of the Telephone*, ed. I. De Sola Pool, 246–261. Cambridge, MA: MIT Press.

Zerubavel, E. 1979. *Patterns of Time in Hospital Life*. Chicago: University of Chicago Press.

Zerubavel, E. 1985. *Hidden Rhythms: Schedules and Calendars in Social Life*. Berkeley: University of California Press.

Zickuhr, K., and A. Smith. 2011. 28% of American adults use mobile and social location-based services. Washington, DC: Pew Internet and American Life Project.

Index

Note: Page numbers with "f" indicate figures; those with "n" indicate endnotes.